Lamborghini Gallardo

A Decade of Domination

Todd Bandel

Copyright © 2024 Todd Bandel

All rights reserved.

ISBN: 9798300933203

DEDICATION

Dedicated to every automotive enthusiast and gearhead who loves performance and power.

CONTENTS

ACKNOWLEDGEMENT   i

CHAPTER ONE
The Birth of a Legend: Lamborghini's Journey to the Gallardo   1

Chapter Two
Concept to Reality: Designing the Gallardo   17

Chapter Three
Under the Hood: The Heart of the Gallardo   33

Chapter Four
Evolution of Power: Engine Developments Through the Years   51

Chapter Five
Shifting Gears: Transmission Innovations in the Gallardo   69

Chapter Six
Form Meets Function: Aerodynamics and Styling Changes   87

Chapter Seven
Special Editions: Limited Run Gallardo Models   105

Chapter Eight
On the Track: The Gallardo's Racing Pedigree   121

Chapter Nine
Technology and Innovation: Electronics and Driver Aids   137

Chapter Ten
The Gallardo's Impact on Lamborghini's Brand and Market Position   153

Chapter Eleven
Comparing Generations: How the Gallardo Evolved Over a Decade   171

Chapter Twelve
Legacy and Influence: The Gallardos' Place in Supercar History   189

ACKNOWLEDGMENTS

I want to express my deepest gratitude to my father for introducing me to the exhilarating world of automotive racing. Your passion for cars and dedication to the sport have inspired me.

From the first time you took me to a race track, I was captivated by the power and precision of the machines, as well as the skill required to master them.

Your guidance and support have fueled my interest and enthusiasm, making every moment in this thrilling world more meaningful. Thank you for sharing this incredible journey with me and being such a pivotal influence in my life.

Chapter 1: The Birth of a Legend: Lamborghini's Journey to the Gallardo

Section 1.1: Ferruccio Lamborghini's Vision

Ferruccio Lamborghini's journey to automotive legend began far from the world of high-performance sports cars. Born in 1916 to grape farmers in the Italian region of Emilia-Romagna, Lamborghini displayed an early aptitude for mechanics that would shape his future. After serving as a vehicle maintenance supervisor in the Italian Royal Air Force during World War II, he returned home with a keen understanding of engines and a surplus of military vehicles.

This surplus became the foundation of Lamborghini's first business venture: repurposing military vehicles into much-needed agricultural tractors. His ingenuity and mechanical prowess quickly established Lamborghini Trattori as a successful tractor-manufacturing company. By the early 1960s, Ferruccio had become one of Italy's most successful industrialists, with interests spanning from air conditioning units to hydraulic valves.

Lamborghini Gallardo: A Decade of Domination

Lamborghini's success afforded him the luxury of indulging in high-performance sports cars, including several Ferraris. However, his experience with these vehicles left him dissatisfied, particularly with their clutches and overall refinement. It was this dissatisfaction that led to the now-famous dispute with Enzo Ferrari.

Legend has it that when Lamborghini approached Ferrari with suggestions for improving their cars, he was dismissively told to stick to building tractors. This exchange, whether apocryphal or not, ignited a fire in Lamborghini. He decided that if Ferrari wouldn't build the perfect grand tourer, he would do it himself.

Ferruccio's vision for his ideal car was clear: it would be a refined grand tourer that combined the performance of a Ferrari with the comfort and reliability of a Mercedes-Benz. He envisioned a vehicle that could be driven at high speeds on the autostrada for hours without fatiguing the driver, yet still turn heads with its style and performance.

In 1963, Automobili Lamborghini was founded in Sant'Agata Bolognese, not far from Ferrari's headquarters in Maranello. Ferruccio's choice of location was no coincidence; it was a direct challenge to the established sports car manufacturer and a statement of his serious intent to compete at the highest level.

Both ambition and challenges marked the early days of Automobili Lamborghini. Ferruccio's experience in manufacturing and his wealth provided a solid foundation, but building a high-performance sports car from scratch was a formidable task. He assembled a team of talented young engineers, including Giotto Bizzarrini, who had previously worked for Ferrari.

One of the first major hurdles was developing an engine that could compete with Ferrari's offerings. Bizzarrini designed a 3.5-liter

V12 engine that would become the heart of Lamborghini's first production car, the 350 GT. This engine, with its impressive power output and smooth operation, was a testament to Lamborghini's commitment to engineering excellence.

The company faced skepticism from the automotive world. Many doubted that a tractor manufacturer could successfully enter the rarified world of high-performance sports cars. However, Ferruccio's determination, combined with the skill of his team, began to turn heads when the first Lamborghini prototype was unveiled at the 1963 Turin Auto Show.

Despite the initial challenges and skepticism, Ferruccio Lamborghini's vision began to take shape. The company's early models, particularly the 350 GT and its successor, the 400 GT, proved that Lamborghini could indeed produce refined, high-performance grand tourers that could compete with the best in the world.

Ferruccio's vision extended beyond just creating fast cars. He wanted to build a brand that represented the pinnacle of Italian automotive engineering and design. This ambition would drive Lamborghini to push boundaries and take risks that would ultimately define the company's DNA and set the stage for future innovations, including the revolutionary Miura and, decades later, the game-changing Gallardo.

Section 1.2: The Era of Iconic Models

Lamborghini's journey to the Gallardo was paved with a series of groundbreaking models that not only defined the brand but also left an indelible mark on the automotive industry. This era of iconic models began with the 350 GT, Lamborghini's first production car, which debuted in 1964. The 350 GT was a testament to Ferruccio Lamborghini's vision of creating a grand tourer that could rival Ferrari.

Lamborghini Gallardo: A Decade of Domination

With its sleek Touring-designed body and powerful V12 engine, the 350 GT immediately established Lamborghini as a serious contender in the high-performance luxury car market.

However, it was the introduction of the Miura in 1966 that truly revolutionized the supercar landscape. The Miura's mid-engine layout was a radical departure from contemporary designs, setting a new standard for high-performance sports cars. Its transversely mounted V12 engine, coupled with its stunning Marcello Gandini-designed bodywork, made the Miura an instant icon. The car's impact on the industry cannot be overstated; it essentially created the modern supercar category and influenced countless designs that followed.

Following the Miura's success, Lamborghini continued to push the boundaries of automotive design with the introduction of the Countach in 1974. The Countach's wedge-shaped silhouette and dramatic scissor doors were a radical departure from traditional sports car design. Its futuristic appearance captured the imagination of car enthusiasts worldwide and came to epitomize the excess and flamboyance of the 1980s.

The Countach's influence on supercar design was profound and long-lasting, inspiring generations of designers and setting a new benchmark for visual drama in automotive styling. As Lamborghini entered the 1990s, the company faced the challenge of modernizing its image while maintaining its reputation for producing extreme, high-performance vehicles. The answer came in the form of Diablo, introduced in 1990. The Diablo retained the dramatic styling cues of its predecessors but incorporated modern technology and improved driveability. It was the first Lamborghini capable of exceeding 200 mph, cementing the brand's position at the pinnacle of automotive performance. The Diablo also saw Lamborghini embrace all-wheel drive technology, a feature that would become increasingly important in future models.

These iconic models played a crucial role in shaping Lamborghini's DNA. They established key characteristics that would come to define the brand: striking, avant-garde design, powerful, high-revving engines, and an uncompromising approach to performance. The 350 GT laid the foundation with its grand tourer ethos and V12 power.

The Miura introduced the mid-engine layout that would become a Lamborghini hallmark. The Countach pushed the boundaries of design and established Lamborghini as a brand unafraid of controversy or excess. Finally, the Diablo showed that Lamborghini could evolve with the times, incorporating modern technology without losing its essential character.

Each of these models contributed something unique to Lamborghini's identity, creating a rich heritage that would inform the development of future cars, including the Gallardo. They set a high bar for performance, design, and emotional appeal, a bar that any new Lamborghini would need to clear.

As the company looked towards the future and the development of a smaller, more accessible model, the legacy of these iconic cars loomed large. The challenge would be to create a car that could live up to this storied lineage while also breaking new ground and appealing to a broader audience. This tension between heritage and innovation is a key factor in the birth of the Gallardo.

Section 1.3: Chrysler's Influence and the Path to Modernization

The 1970s and 1980s were tumultuous times for Lamborghini, as the company faced significant financial challenges that threatened its very existence. The 1973 oil crisis hit the automotive industry hard, and luxury sports car manufacturers like Lamborghini were particularly vulnerable. Sales plummeted, and the company struggled to stay afloat amid mounting debts and dwindling resources.

In 1987, a turning point arrived when Chrysler Corporation, under the leadership of Lee Iacocca, acquired Lamborghini. This acquisition marked a new era for the Italian supercar maker, bringing American management practices and financial stability to the beleaguered company. Chrysler's involvement was more than just a financial lifeline; it represented a shift in Lamborghini's approach to business and car development.

The impact of American management on Lamborghini's operations was significant and multifaceted. Chrysler introduced more rigorous financial controls and streamlined production processes, helping to stabilize the company's finances. The American auto giant also brought a fresh perspective on market positioning and brand management, encouraging Lamborghini to broaden its appeal while maintaining its exclusive image.

Under Chrysler's ownership, Lamborghini began developing new models that would carry the brand into the 1990s and beyond. The Diablo, which succeeded the iconic Countach, was conceived and developed during this period. This new flagship model showcased a blend of Lamborghini's traditional styling cues with more modern engineering and technology, much of which was influenced by Chrysler's resources and expertise.

Perhaps most significantly, it was during the Chrysler era that the seeds of the "baby Lamborghini" concept were first planted. Recognizing the potential for a more accessible model to complement the extreme, high-performance flagships, Chrysler executives and Lamborghini engineers began exploring the idea of a smaller, more affordable Lamborghini that could compete with the likes of Ferrari's lower-end models and Porsche's 911.

This concept, while not fully realized under Chrysler's ownership, is a prescient move. It laid the groundwork for what would eventually become the Gallardo, although the full fruition of this idea would have to wait for the next chapter in Lamborghini's corporate history.

The Chrysler years, while relatively short-lived, played a crucial role in modernizing Lamborghini and setting the stage for its future success. By introducing more robust business practices, initiating the development of new models, and planting the seed for a broader product range, Chrysler helped transform Lamborghini from a niche, financially unstable manufacturer into a company poised for growth and innovation in the supercar market.

As Lamborghini entered the 1990s, it was a company reborn, still distinctly Italian in its passion and design flair, but now backed by the resources and business acumen of a major American automaker. This unique blend of Italian artistry and American pragmatism is a powerful combination, setting the stage for the next phase of Lamborghini's evolution and, ultimately, the birth of the Gallardo.

Section 1.4: The Audi Era Begins

The year 1998 marked a pivotal moment in Lamborghini's history as the Volkswagen Group acquired the iconic Italian supercar manufacturer. This acquisition ushered in a new era for Lamborghini, one that would be characterized by German precision, engineering prowess, and a renewed focus on quality and reliability.

Under the Volkswagen Group's umbrella, Audi was tasked with managing Lamborghini. This decision was strategic, as Audi's reputation for advanced technology and high-quality engineering aligned well with Lamborghini's aspirations. The marriage of Italian passion and German engineering precision promised to elevate Lamborghini to new heights.

Lamborghini Gallardo: A Decade of Domination

The influx of German engineering and quality control practices had a profound impact on Lamborghini's operations. Audi's influence brought about significant improvements in manufacturing processes, quality assurance, and overall vehicle reliability. This was a crucial development for Lamborghini, which had historically struggled with consistency and dependability issues. The implementation of rigorous German quality standards helped address these concerns, enhancing the brand's reputation and customer satisfaction.

With Audi's backing, Lamborghini began to expand its model range strategy. The company recognized the need to diversify its lineup beyond the flagship V12 models that had been its mainstay for decades. This strategic shift aimed to broaden Lamborghini's appeal and increase its market share in the high-performance luxury segment.

Perhaps the most significant decision of this era was the commitment to develop a smaller, more accessible Lamborghini. This concept, which would eventually evolve into the Gallardo, was a departure from Lamborghini's traditional focus on extreme, limited-production supercars. The idea was to create a model that could compete with the likes of Ferrari's lower-tier offerings and Porsche's 911, opening up Lamborghini ownership to a broader audience without diluting the brand's exclusivity and performance credentials.

The decision to pursue this "baby Lamborghini" project was not without controversy. Some purists worried that a more attainable model might diminish Lamborghini's ultra-exclusive image. However, Audi's management, along with Lamborghini's leadership, recognized the potential for significant growth and long-term sustainability that such a model could bring.

This period of Audi stewardship set the stage for Lamborghini's renaissance. By combining German engineering expertise with Italian design flair and passion, Lamborghini was poised to enter a new phase of its existence. The groundwork laid during this time would prove instrumental in the development and success of the Gallardo. This car would go on to become the best-selling Lamborghini in history and redefine the brand for the 21st century.

Section 1.5: Market Analysis and the Need for the Gallardo

As the new millennium approached, the supercar market was experiencing a renaissance. Established players like Ferrari and Porsche were pushing the boundaries of performance and luxury, while new entrants vied for a slice of this lucrative market. Lamborghini, despite its storied history and iconic models, found itself at a crossroads. The company's leadership recognized the need for a strategic move to secure its future in this evolving landscape.

The late 1990s and early 2000s saw a shift in consumer preferences within the high-performance car segment. While there was still a demand for the ultimate, no-compromise supercars like Lamborghini's Diablo, a new market was emerging. Wealthy enthusiasts were increasingly seeking vehicles that offered supercar performance but with greater usability for everyday driving. This trend was exemplified by the success of models like the Ferrari 360 Modena and the Porsche 911 Turbo.

Lamborghini's existing lineup, centered around the aging Diablo, left a significant gap in its offerings. The brand lacked a model that could compete directly with these more "accessible" supercars. This gap not only limited Lamborghini's potential market share but also restricted its ability to attract a new generation of buyers who might eventually graduate to their flagship models.

Lamborghini Gallardo: A Decade of Domination

A comprehensive competitor analysis revealed the strengths and weaknesses of Lamborghini's rivals. Ferrari's 360 Modena was praised for its blend of performance and relative ease of use.

Porsche's 911 series continued to set the benchmark for everyday usability in a high-performance package. Meanwhile, emerging brands like Pagani were carving out niches in the ultra-high-end segment, threatening Lamborghini's position as the ultimate expression of automotive excess.

The Lamborghini team identified a clear opportunity: a model that could offer the brand's signature flamboyance and performance, but in a more compact and user-friendly package. This new car would need to embody Lamborghini's DNA while appealing to a broader audience. It would serve as an entry point to the brand, potentially doubling or even tripling Lamborghini's annual sales figures.

The target customer for this new model was meticulously profiled. They envisioned successful buyers, passionate about cars, and sought the prestige of the Lamborghini brand. However, these customers also valued practicality and the ability to use their supercar more frequently than a traditional Lamborghini might allow. The ideal buyer would be slightly younger than the typical Diablo customer, possibly making their first foray into the supercar market.

Projections for the impact of this new model on Lamborghini's sales and brand positioning were ambitious. The company anticipated that a successful "baby Lamborghini" could potentially double their annual sales volume. Moreover, it would position Lamborghini as a more comprehensive luxury sports car manufacturer, capable of competing across multiple segments of the high-performance market.

The potential risks were also carefully considered. There were concerns about diluting the brand's exclusive image and the possibility of cannibalizing sales from the higher-end models. However, the leadership team believed that the benefits far outweighed the risks. They saw this new model as essential for ensuring Lamborghini's long-term viability and growth in an increasingly competitive market.

As the concept for what would become the Gallardo began to take shape, excitement built within Lamborghini and the broader automotive industry. The stage was set for a car that would not only fill a gap in Lamborghini's lineup but potentially redefine the entire brand for a new era.

Section 1.6: The Gallardo Project Takes Shape

As the new millennium dawned, Lamborghini stood at a critical juncture. With Audi's backing and a clear vision for the future, the company was ready to embark on its most ambitious project yet: the development of the Gallardo. This new model would not only fill a gap in Lamborghini's lineup but also redefine the brand for a new era.

The formation of the Gallardo development team marked the beginning of an exciting new chapter for Lamborghini. Drawing from the best talent within the company and leveraging expertise from Audi, the team was a blend of Italian passion and German precision. At its helm was Maurizio Reggiani, Lamborghini's newly appointed Head of R&D, whose task was to bridge the gap between Lamborghini's storied heritage and the demands of modern automotive engineering.

The initial design briefs for the Gallardo were both exhilarating and daunting. The car needed to embody the essence of Lamborghini: dramatic styling, blistering performance, and an unmistakable presence on the road. Yet it also had to be more accessible, both in terms of drivability and market positioning. This delicate balance would define the entire development process.

Engineering goals for the Gallardo were equally ambitious. The team aimed to create a car that could outperform its rivals in acceleration, handling, and overall driving dynamics. At the same time, it needed to offer a level of reliability and everyday usability that had often eluded previous Lamborghini models. This meant incorporating cutting-edge technology while maintaining the raw, emotional appeal that Lamborghini was known for.

Balancing Lamborghini's heritage with modern technology became a central theme of the Gallardo project. The team looked to iconic models like the Miura and Countach for inspiration, seeking to capture their spirit in a thoroughly modern package. This meant retaining signature elements like the mid-engine layout and dramatic styling while integrating contemporary features such as advanced electronics and driver aids.

The collaboration between Italian and German engineers proved to be a key factor in Gallardo's development. Lamborghini's team brought their expertise in creating emotionally resonant, high-performance machines, while Audi's engineers contributed their knowledge of advanced manufacturing techniques, quality control, and systems integration. This cross-cultural exchange led to heated debates but ultimately resulted in a synthesis of ideas that would elevate Gallardo above its competitors.

Key decisions were made in powertrain, chassis, and overall vehicle architecture to create a true Lamborghini that could also serve as a daily driver. The team settled on a V10 engine, a first for Lamborghini, which offered an ideal balance of power, weight, and packaging. The chassis was designed to be both rigid and lightweight, utilizing aluminum and other advanced materials. All-wheel drive was chosen to enhance stability and traction, a departure from Lamborghini's traditional rear-wheel-drive layout.

As the Gallardo project progressed, excitement within Lamborghini grew palpable. Engineers and designers worked tirelessly, often pushing the boundaries of what was thought possible. Every component, from the aerodynamic body panels to the bespoke interior trim, was scrutinized and refined. The goal was not just to create another supercar but to craft a machine that would captivate a new generation of enthusiasts while staying true to Lamborghini's core values.

Throughout the development process, the team faced numerous challenges. Integrating advanced electronics with Lamborghini's traditionally analog approach to driving dynamics required innovative solutions. Ensuring that the car could meet stringent safety and emissions regulations without compromising performance demanded creative engineering. Yet with each obstacle overcome, the Gallardo inched closer to reality.

As the project neared completion, anticipation within the automotive world reached a fever pitch. Spy photos and rumors circulated, fueling speculation about Lamborghini's new creation. Inside the company, there was a growing sense that they had created something truly special - a car that would not only meet but exceed the lofty expectations placed upon it.

The Gallardo project represented more than just the development of a new model; it was the rebirth of Lamborghini for the 21st century. As the team put the finishing touches on their creation, they knew they had shaped not just a car, but the future of the brand itself. The stage was set for the Gallardo to make its mark on automotive history.

Section 1.7: From Concept to Production

The journey of the Lamborghini Gallardo from initial concept to production-ready supercar was a testament to the brand's commitment to innovation and excellence. As the project gained momentum, the development team, a blend of Italian passion and German precision, worked tirelessly to bring their vision to life.

The Gallardo's concept phase was a crucible of creativity and engineering prowess. Designers and engineers collaborated closely, sketching, modeling, and refining their ideas. They sought to create a car that was unmistakably Lamborghini, yet accessible to a broader audience. The concept underwent several iterations, each refining the car's lines and proportions to achieve the perfect balance of aggression and elegance.

As the concept solidified, the team moved into the prototype development stage. This phase was crucial in translating the stunning designs on paper into a functioning, high-performance machine. Multiple prototypes were built, each serving as a testbed for different aspects of the car's performance and design. These early prototypes underwent rigorous testing, from wind tunnel sessions to evaluate aerodynamics, to high-speed runs on Lamborghini's test track in Sant'Agata Bolognese.

The development process was not without its challenges. One of the most significant hurdles was integrating Audi's all-wheel-drive system into the Gallardo while maintaining Lamborghini's characteristic handling dynamics. The engineering team worked tirelessly to fine-tune the system, ensuring it enhanced rather than detracted from the car's performance. Another challenge was achieving the perfect balance between everyday usability and supercar performance, a key goal for the Gallardo project.

As the prototypes evolved, so did the excitement within Lamborghini. Each milestone achieved brought the team closer to their goal of creating a game-changing supercar. The final design approval was a momentous occasion, signaling the transition from development to production preparation. The Sant'Agata Bolognese factory underwent significant upgrades to accommodate the new model, with state-of-the-art production lines installed to ensure each Gallardo would be built to the highest standards.

In the months leading up to the Gallardo's unveiling, anticipation within the automotive world reached a fever pitch. Spy photos of camouflaged prototypes fueled speculation about the car's capabilities and design. Lamborghini carefully orchestrated a teaser campaign, releasing tantalizing glimpses of the vehicle to whet the appetite of enthusiasts and potential buyers alike.

As the launch date approached, the entire Lamborghini team felt a mix of excitement and nervous energy. They knew they had created something special, a car that would not only live up to the Lamborghini name but also usher in a new era for the brand. The Gallardo represented years of hard work, innovation, and passion, all distilled into a compact, high-performance package.

With final quality checks complete and the production line primed, Lamborghini stood ready to introduce the world to the Gallardo. The stage was set for a launch that would send shockwaves through the automotive industry and cement Lamborghini's position as a leader in the supercar market. The birth of a legend was imminent, and the automotive world waited with bated breath to witness the unveiling of the car that would redefine Lamborghini for the 21st century.

Lamborghini Gallardo: A Decade of Domination

Chapter 2: Concept to Reality: Designing the Gallardo

Section 2.1: Initial Concept and Design Brief

The journey of the Lamborghini Gallardo began with a bold vision to create a more accessible supercar that would complement the flagship Murciélago while maintaining the brand's DNA. In the early 2000s, under Audi's ownership, Lamborghini set out to increase its production volumes without compromising the exclusivity that had long been a hallmark of the brand. The Gallardo was conceived as an entry-level Lamborghini that would broaden the company's appeal and reach a wider audience of supercar enthusiasts.

To bring this vision to life, Lamborghini assembled a world-class design team led by Luc Donckerwolke, the company's head of design at the time. Donckerwolke, known for his innovative approach and keen eye for detail, was tasked with creating a car that would be unmistakably Lamborghini, yet more compact and usable than its larger sibling, the Murciélago. The design team also included talented individuals from both Lamborghini and parent company Audi, bringing together a wealth of experience and fresh perspectives.

The design brief for the Gallardo was both exciting and challenging. It called for a car that would embody the essence of Lamborghini - dramatic, aggressive, and uncompromisingly exotic - while also introducing a new level of everyday usability to the brand. The team was given clear parameters: the car needed to be smaller than the Murciélago, more accessible in terms of driving dynamics, and capable of being produced in higher numbers without losing the hand-crafted feel that Lamborghini was known for.

Drawing inspiration from Lamborghini's rich heritage, the design team looked to iconic models like the Miura and Countach for guidance. These legendary cars had defined the supercar genre in their respective eras, and the Gallardo needed to carry that torch into the 21st century. Elements such as the low-slung profile, aggressive wedge shape, and distinctive air intakes were reimagined and reinterpreted for a new generation of Lamborghini enthusiasts.

One of the most significant challenges faced by the design team was striking a balance between form and function. Every curve, every angle, and every surface of the Gallardo had to serve both aesthetic and aerodynamic purposes. The car needed to be visually stunning, instantly recognizable as a Lamborghini, and capable of turning heads wherever it went. At the same time, it had to deliver the exceptional performance that buyers expected from a car bearing the raging bull emblem.

This delicate balance was achieved through countless hours of sketching, modeling, and refining. The design team explored various concepts, ranging from radical departures to more evolutionary designs, while always keeping the core values of the Lamborghini brand in mind. They worked closely with engineers to ensure that every design decision supported the car's performance goals, from optimizing airflow to reducing weight.

The result of this intensive process was a design that managed to be both fresh and familiar. The Gallardo's lines were clean and modern, yet unmistakably Lamborghini. Its compact dimensions gave it a sense of agility and purpose, while clever design elements, such as the distinctive side air intakes and sharp creases, added drama and visual interest.

As the initial concept began to take shape, it became clear that the Gallardo would be more than just another supercar. It represented a new chapter for Lamborghini, a bold step into a future where performance and practicality could coexist. The design brief had set the stage, and now it was time to transform this exciting concept into a tangible reality.

Section 2.2: The Design Process

The journey from concept to reality for the Lamborghini Gallardo was a meticulous and iterative process that combined artistic vision with cutting-edge technology. This section explores the various stages of the design process that brought the Gallardo to life.

The design process began with initial sketches and 2D renderings. Designers at Lamborghini explored a wide range of concepts, from radical departures that pushed the boundaries of the brand's design language to more evolutionary designs that built upon the company's rich heritage. These early sketches were crucial in establishing the overall direction of Gallardo's design, allowing the team to experiment with different proportions, lines, and details. The most promising concepts were then refined and developed further.

As the design progressed, the team transitioned from 2D to 3D modeling and computer-aided design (CAD). Advanced CAD software allowed designers to refine the Gallardo's shape with millimeter precision, ensuring that every curve and angle served both aesthetic and functional purposes.

This digital modeling phase was instrumental in visualizing how different design elements would work together and how they would translate to a three-dimensional form. It also enabled early aerodynamic simulations, allowing for the identification of potential issues before physical prototypes were built.

Despite the advantages of digital design tools, the importance of physical models in the design process cannot be overstated. Clay modeling and full-scale mock-ups played a crucial role in the Gallardo's development. The full-scale clay model allowed designers to evaluate proportions and make real-time adjustments that might not have been apparent in digital renderings.

This hands-on approach enabled the design team to fine-tune details such as surface transitions, panel gaps, and overall stance. The clay model also served as a valuable tool for presenting the design to Lamborghini's management and gathering feedback.

With the basic shape established, the design team moved on to wind tunnel testing and aerodynamic refinement. Extensive wind tunnel testing led to subtle adjustments that significantly improved downforce and stability without compromising the car's striking aesthetics. This process was crucial in ensuring that the Gallardo would not only look fast but also perform exceptionally at high speeds. The team worked tirelessly to optimize airflow around the car, paying particular attention to areas such as the front splitter, side skirts, and rear diffuser.

While the exterior design was being refined, equal attention was given to the interior design and ergonomics. The Gallardo's cabin needed to balance luxury with a driver-focused layout, embodying Lamborghini's commitment to creating ultimate driving machines. Designers carefully considered every aspect of the interior, from the placement of controls to the quality of materials used. The goal was to create a cockpit that would immerse the driver in the driving experience while providing the comfort and refinement expected of a high-end supercar.

Throughout the design process, a constant dialogue was maintained between designers and engineers to ensure that aesthetic decisions aligned with technical requirements. This collaborative approach was essential in creating a car that was not only visually stunning but also a technological marvel.

The design process of the Lamborghini Gallardo was a testament to the brand's commitment to excellence. It showcased how modern design tools and traditional craftsmanship could be combined to create a truly exceptional automobile. The result was a car that perfectly balanced form and function, setting a new standard for supercar design in the 21st century.

Section 2.3: Engineering Challenges and Solutions

The journey from concept to reality for the Lamborghini Gallardo was fraught with engineering challenges that required innovative solutions. As the design team worked to create a visually stunning supercar, they had to balance aesthetics with the practical demands of high-performance engineering constantly.

One of the primary challenges was integrating the powertrain into the design. The Gallardo's compact V10 engine presented both opportunities and obstacles. Its relatively small size allowed for a lower hood line, enhancing both the car's sleek aesthetics and forward visibility for the driver.

However, packaging such a powerful engine in a compact space required meticulous planning. Engineers worked closely with designers to ensure that every component fit perfectly while maintaining the car's striking silhouette. The result was a harmonious blend of form and function, with the engine becoming an integral part of the car's overall design.

The chassis and suspension design posed another significant challenge. The team needed to create a structure that could handle the immense power of the V10 engine while providing the agility and responsiveness expected of a Lamborghini. The solution came in the form of an all-new aluminum spaceframe, designed to provide both rigidity and lightness.

This innovative approach allowed the Gallardo to achieve exceptional handling characteristics without compromising the purity of its design. The suspension system was carefully tuned to work in concert with the chassis, ensuring that the car remained stable and composed even under extreme driving conditions.

Cooling and airflow management presented a unique set of challenges. High-performance engines generate significant heat, which needs to be efficiently dissipated to maintain optimal performance. However, traditional large air intakes and vents could disrupt the clean lines of the Gallardo's design. The engineering team devised an innovative solution by integrating air intakes into the body panels themselves. These cleverly disguised openings provided efficient cooling without compromising the car's aesthetics. The side intakes, for instance, were designed to channel air directly to the engine and brakes, while maintaining the smooth, unbroken surface of the car's flanks.

The use of lightweight materials and advanced construction techniques was crucial in achieving the performance targets while minimizing weight. Extensive use of aluminum and carbon fiber throughout the car's structure and body panels helped reduce overall mass without sacrificing strength or rigidity.

This approach not only improved performance but also enhanced fuel efficiency and handling. The challenge lay in incorporating these materials in a way that was both structurally sound and visually appealing. The engineering team worked tirelessly to ensure that every component, from the carbon fiber rear diffuser to the aluminum body panels, contributed to both the car's performance and its stunning appearance.

One of the most significant challenges was adapting the design for series production. While concept cars can be built with little regard for manufacturing practicalities, a production model must be reproducible with consistent quality.

This meant that certain design elements had to be modified to ensure they could be reliably manufactured on a larger scale. For example, complex curves that looked stunning on a concept model might need to be simplified slightly to allow for consistent panel gaps in production. The team had to find creative solutions to maintain the essence of the original design while making it feasible for mass production.

Throughout this process, the engineering team worked closely with the design team, constantly communicating and compromising to achieve the best possible outcome. Every decision, from the placement of a cooling vent to the shape of a suspension component, had to be carefully considered in terms of both its functional impact and its effect on the overall design.

The result of these efforts was a car that not only looked spectacular but also performed exceptionally well. The Gallardo's engineering solutions allowed it to achieve supercar performance while maintaining a level of usability and reliability that was unprecedented in its class. This successful integration of advanced engineering with stunning design set a new standard for Lamborghini and influenced the development of supercars for years to come.

Section 2.4: Design Evolution and Refinement

A continuous process of evolution and refinement marked the journey from concept to production for the Lamborghini Gallardo. This iterative approach ensured that the final product not only met the high standards set by Lamborghini but also exceeded the expectations of its discerning clientele. A crucial element in this refinement process was the feedback gathered from focus groups and potential customers. Lamborghini understood that while its designers and engineers were experts in their fields, the ultimate judges of the Gallardo's appeal would be its target market.

High-net-worth individuals were invited to exclusive preview events where they could examine design mock-ups and provide their insights. This valuable feedback led to subtle yet significant refinements in the car's proportions and overall aesthetic. For instance, the rake of the windshield was slightly adjusted based on customer preferences, striking a perfect balance between aggressive styling and practical visibility.

Equally important was the input from racing drivers and performance experts. Lamborghini's commitment to creating a valid driver's car meant that the Gallardo's design had to be both functional and beautiful. Test drivers pushed prototypes to their limits on both track and road, providing crucial feedback that shaped the car's final form.

One notable example was the redesign of the rear diffuser following extensive high-speed testing. The original design was found to create slight instability at extreme speeds, prompting aerodynamicists to revise the shape for improved downforce without compromising the car's sleek rear-end styling.

Regulatory compliance and safety considerations played a significant role in the design evolution as well. The challenge was to meet increasingly stringent safety standards without diluting the Gallardo's dramatic design. The front bumper, in particular, underwent several iterations to satisfy pedestrian safety requirements while maintaining the car's aggressive face. Clever use of materials and innovative structural designs allowed Lamborghini to create a front end that was both safe and visually striking.

The design process was highly iterative, with multiple prototypes built and tested. Each version brought new insights and challenges, prompting the team to refine their work continually. Advanced computer simulations were complemented by real-world testing, ensuring that every curve and contour served both form and function. This meticulous approach extended to even the most minor details, such as the design of the side mirrors, which were sculpted to minimize wind noise while maintaining the car's overall aesthetic flow.

As the design neared completion, the focus shifted to preparing for production. This stage involved close collaboration between designers, engineers, and manufacturing experts to ensure that the Gallardo's complex shapes and features could be consistently reproduced on the assembly line. Some design elements required innovative production techniques, such as the development of new tooling processes to create the car's distinctive, sharp creases and panels.

The final design approval was a momentous occasion, marking the culmination of years of work. Lamborghini's board conducted an exhaustive review of every design aspect, from the exterior styling to the interior ergonomics. This rigorous evaluation ensured that the Gallardo not only met but exceeded the original brief, embodying the perfect blend of Italian passion and German engineering precision that defined the new era of Lamborghini under Audi ownership.

With approval secured, the team moved swiftly into pre-production preparations. This phase involved finalizing tooling, establishing quality control processes, and training assembly teams. The goal was to ensure that every Gallardo that rolled off the production line in Sant'Agata Bolognese would be a perfect realization of the designers' vision.

The evolution and refinement of the Gallardo's design were a testament to Lamborghini's commitment to excellence. It demonstrated that creating a truly exceptional supercar requires not only initial creativity but also the patience and dedication to improve and adapt continually. The result was a car that not only captivated audiences at its unveiling but continued to turn heads and ignite passions throughout its decade-long production run, cementing its place as one of the most iconic designs in Lamborghini's storied history.

Section 2.5: Design Features of the Production Gallardo

The production Lamborghini Gallardo was a masterpiece of automotive design, seamlessly blending form and function to create a truly iconic supercar. Its exterior styling elements were a testament to the brand's commitment to bold, aggressive aesthetics. The car's low-slung profile and sharply angled lines gave it a sense of motion even when stationary. At the same time, the distinctive Y-shaped LED daytime running lights became a signature Lamborghini design element that would influence future models.

Aerodynamics played a crucial role in shaping the Gallardo's final form. Every curve and contour was meticulously crafted to optimize airflow and enhance performance. The integrated rear spoiler, for instance, was a marvel of engineering and design. Deployed automatically at high speeds, it increased downforce without disrupting the car's sleek silhouette when retracted. This attention to aerodynamic detail not only improved the car's performance but also contributed to its striking visual appeal.

The Gallardo's design was replete with unique Lamborghini cues that firmly established its lineage. The hexagonal theme, seen in the side air intakes and rear lights, became a recurring motif in Lamborghini designs, creating a visual language that would define the brand for years to come. These distinctive elements ensured that the Gallardo was instantly recognizable as a Lamborghini, even from a distance.

Color played a significant role in bringing the Gallardo's design to life. The range of exterior finishes offered was carefully curated to accentuate the car's lines and surfaces. Perhaps none was more impactful than the iconic Verde Ithaca green, which became synonymous with the Gallardo. This vibrant hue highlighted the car's dramatic angles and became a favorite among enthusiasts, further cementing the Gallardo's place in automotive design history.

The interior of the Gallardo was just as meticulously designed as its exterior. The cabin was a perfect blend of luxury and sportiness, with a clear focus on the driver's experience. The cockpit featured a distinctive instrument cluster inspired by fighter jets, placing crucial information directly in the driver's line of sight. High-quality materials such as fine leather, aluminum, and carbon fiber were used throughout, creating an environment that was both opulent and performance-oriented.

Every aspect of the Gallardo's interior was designed to enhance the driving experience. The seats were sculpted to provide both comfort and support during high-speed maneuvers. The center console was angled towards the driver, placing all controls within easy reach. Even the steering wheel was a work of art, its shape and size carefully optimized for both power and aesthetics.

The Gallardo's design was a triumph of Italian styling and German engineering precision. It successfully walked the fine line between maintaining Lamborghini's flamboyant design heritage and introducing a more refined, sophisticated aesthetic. The car's compact yet muscular proportions, aggressive stance, and carefully considered details all came together to create a design that was both timeless and groundbreaking.

Perhaps most importantly, the Gallardo's design achieved its goal of making Lamborghini more accessible without diluting the brand's core values. It was unmistakably a Lamborghini, yet it had a level of usability and everyday practicality that was unprecedented for the marque. This balance of extreme performance and relative user-friendliness was reflected in every aspect of its design.

In essence, the design features of the production Gallardo represented a new chapter in Lamborghini's design philosophy. It set new standards not just for the brand, but for the entire supercar segment. The Gallardo's design proved that a car could be both outrageously exotic and surprisingly usable, a combination that would influence automotive design for years to come.

Section 2.6: Design Legacy and Impact

The Lamborghini Gallardo's design left an indelible mark on both the automotive industry and the Lamborghini brand itself. Its striking aesthetics and innovative features set a new standard for supercar design, influencing not only future Lamborghini models but also competitors across the luxury sports car segment.

One of the most significant impacts of the Gallardo's design was its role in modernizing Lamborghini's visual language. The car's crisp lines and angular form represented a departure from the more curvaceous designs of the past, setting the tone for a new era of Lamborghini aesthetics. This design philosophy would go on to influence subsequent models, including the Huracán and Aventador, creating a cohesive and instantly recognizable family look for the brand.

The Gallardo's compact yet aggressive stance also challenged prevailing notions of supercar proportions. By proving that a smaller footprint could still deliver a commanding presence, it opened up new possibilities for designers across the industry. This shift in thinking paved the way for a new generation of more accessible supercars, blending everyday usability with exotic performance and styling.

One of the most enduring legacies of the Gallardo's design was its emphasis on aerodynamic integration. Rather than relying on obvious spoilers and wings, the car's body was sculpted to manage airflow efficiently while maintaining clean lines. This approach to "functional beauty" became a hallmark of modern supercar design, with many manufacturers following suit in subsequent years.

The Gallardo's interior design also left a lasting impression on the automotive world. Its driver-focused cockpit, featuring a distinctive instrument cluster and center console, sets a new standard for ergonomics in high-performance vehicles. The blend of luxury materials with a purposeful, almost minimalist layout influenced not only future Lamborghinis but also sparked a trend across the broader sports car market.

Color played a crucial role in the Gallardo's design impact as well. The introduction of bold, vibrant hues like Verde Ithaca green and Arancio Borealis orange became synonymous with the m, reinvigorating the use of daring colors in the luxury car segment. This color strategy helped to differentiate Lamborghini from its more conservative competitors further and became a key part of the brand's exuberant image.

The design's versatility was another aspect that contributed to its lasting impact. Over its decade-long production run, the Gallardo's basic shape proved remarkably adaptable, spawning numerous special editions and variants without losing its core identity. This ability to evolve while maintaining brand consistency became a valuable lesson for automotive designers worldwide.

Notably, the Gallardo's design also played a crucial role in Lamborghini's commercial success. Its appealing aesthetics, combined with its more accessible positioning, helped significantly boost the brand's sales volumes. This commercial triumph demonstrated the power of thoughtful design in driving business success, influencing how other luxury brands approached their product strategies.

The car's design also had a cultural impact beyond the automotive world. Its distinctive silhouette became instantly recognizable, featuring prominently in popular media, from movies and video games to art and fashion. This cultural penetration helped to cement Lamborghini's status as a global icon of automotive design and performance. In conclusion, the Gallardo's design legacy extends far beyond its decade-long production run.

It represented a pivotal moment in Lamborghini's history, setting new standards for the brand and the wider automotive industry. The principles of balanced aggression, functional beauty, and adaptable design that were embodied in the Gallardo continue to influence supercar aesthetics to this day, cementing its place as a true design icon of the 21st century.

Section 2.7: Legacy and Influence of the Gallardo Design

The Lamborghini Gallardo's design left an indelible mark on the automotive world, influencing not only future Lamborghini models but also the broader supercar industry. Its sleek, angular form became instantly recognizable and set a new standard for what a modern supercar should look like.

One of the most significant aspects of the Gallardo's design legacy was its ability to balance aggression and elegance. The car's sharp lines and low-slung profile exuded the traditional Lamborghini ferocity. At the same time, its compact dimensions and refined details added a level of sophistication previously unseen in the brand's lineup. This combination proved irresistible to buyers and critics alike, helping to expand Lamborghini's appeal beyond its traditional customer base.

The Gallardo's design also played a crucial role in establishing new visual cues that would become hallmarks of Lamborghini's design language. The distinctive Y-shaped LED daytime running lights, for instance, became a signature element that would be incorporated into subsequent models. Similarly, the hexagonal theme seen in various aspects of the Gallardo's design, from its side air intakes to its rear lights, evolved into a recurring motif in future Lamborghini vehicles.

Perhaps most importantly, the Gallardo's design demonstrated that a more compact and accessible supercar could still embody the essence of Lamborghini. This concept paved the way for the brand to explore new market segments without compromising its core values. The success of the Gallardo's design approach directly influenced the development of its successor, the Huracán, as well as other models in the Lamborghini range.

Beyond Lamborghini, the Gallardo's design had a ripple effect throughout the automotive industry. Its sharp, geometric forms and aggressive stance inspired numerous competitors and even influenced the design of more mainstream sports cars. The Gallardo proved that a modern supercar could be both exotic and relatively practical, a concept that many manufacturers have since embraced.

The longevity of the Gallardo's design is a testament to its success. Over its decade-long production run, the car received only minor styling updates, with its core design remaining essentially unchanged. This enduring appeal is rare in the fast-paced world of automotive design, where models are often significantly restyled every few years.

In the world of car design, few vehicles manage to become truly iconic. The Lamborghini Gallardo, with its perfect blend of drama and refinement, achieved this rare status. Its design not only fulfilled its immediate goals of expanding Lamborghini's market presence but also set a new direction for the brand and the industry as a whole. The Gallardo's form will continue to be admired and studied by automotive enthusiasts and designers for generations to come, cementing its place in the pantheon of great car designs.

Chapter 3: Under the Hood: The Heart of the Gallardo

Section 3.1: The Genesis of the Gallardo's V10

The decision to develop a V10 engine for the Lamborghini Gallardo marked a significant departure from the company's traditional V12 powerplants. This strategic choice was driven by several factors, chief among them the desire to create a more compact and efficient engine that could still deliver the performance expected of a Lamborghini. The V10 configuration offered an ideal balance between the relative compactness of a V8 and the raw power of a V12, allowing Lamborghini to create a supercar that was both thrilling to drive and more accessible than its larger siblings.

Lamborghini's then-recent acquisition by Audi greatly influenced the development of the Gallardo's V10. This partnership brought together Italian passion for high-performance cars with German engineering precision, resulting in an engine that combined the best of both worlds. The collaboration allowed Lamborghini to leverage Audi's advanced technologies and manufacturing processes, while still maintaining its unique character and performance DNA. This shared development not only resulted in significant cost efficiencies but also led to technological advancements that would benefit both brands in the years to come.

When the Gallardo was first introduced in 2003, its 5.0-liter V10 engine was a masterpiece of engineering. The initial specifications were impressive: 493 horsepower and 376 lb-ft of torque, numbers that put the Gallardo firmly in supercar territory. These figures were achieved through a combination of advanced technologies and meticulous engineering, setting a new standard for performance in its class.

One of the most innovative features of the Gallardo's V10 was its dry-sump lubrication system. This racing-derived technology allowed the engine to be mounted lower in the chassis, significantly lowering the car's center of gravity. The result was enhanced handling and stability, crucial factors in a high-performance supercar. Additionally, the dry-sump system ensured consistent oil pressure even under extreme cornering forces, a common occurrence in spirited driving or on the racetrack.

The impact of this powerful and innovative engine on the Gallardo's performance was profound. The V10's impressive power-to-weight ratio enabled the car to accelerate from 0 to 60 mph in just 4.2 seconds, a blistering pace that put it among the fastest cars of its time. But it wasn't just about straight-line speed; the engine's responsiveness and linear power delivery gave the Gallardo a driving character that was both exhilarating and manageable.

Perhaps most importantly, the V10 engine gave the Gallardo a unique voice in the supercar world. Its distinctive exhaust note, a harmonious blend of high-pitched wail and deep growl, became one of the car's most celebrated features. This auditory signature helped establish the Gallardo's identity, setting it apart from its V8 and V12-powered rivals.

The creation of the Gallardo's V10 engine was more than just a technical achievement; it was a statement of intent from Lamborghini. It signaled the brand's ability to innovate and adapt while staying true to its heritage of building some of the world's most exciting sports cars. This engine would go on to form the foundation of Lamborghini's future, influencing the development of subsequent models and cementing the V10 configuration as a cornerstone of the brand's identity in the 21st century.

Section 3.2: Engine Architecture and Design

The heart of the Lamborghini Gallardo, its formidable V10 engine, was a masterpiece of engineering that set new standards in the supercar world. At its core was a block design that exemplified the perfect balance between strength and weight.

Crafted from aluminum alloy, this lightweight yet robust foundation contributed significantly to the Gallardo's impressive power-to-weight ratio. The use of aluminum not only reduced overall mass but also improved heat dissipation, a crucial factor in maintaining optimal performance under the extreme conditions often experienced by supercars.

The cylinder head configuration of the Gallardo's V10 was equally impressive. Featuring a four-valve-per-cylinder design, it optimized both airflow and combustion efficiency. This setup was further enhanced by a variable valve timing system, which allowed for precise control over valve operation across the entire rev range. As a result, the engine delivered an optimal balance of low-end torque for everyday drivability and high-end power for those exhilarating moments on the track or open road. This versatility was a key factor in making the Gallardo not just a weekend toy, but a supercar that could be enjoyed in a variety of driving scenarios.

The engine's breathing apparatus was another area where Lamborghini's engineers showcased their expertise. The intake and exhaust systems were meticulously designed to maximize airflow while also crafting the signature Lamborghini sound. The tuned intake manifold, in particular, was a work of art that enhanced both performance and the engine's aural character.

It was carefully engineered to provide the optimal air-fuel mixture to each cylinder, ensuring maximum power output across the rev range. The exhaust system, with its precisely calculated tube lengths and diameters, not only helped in scavenging exhaust gases efficiently but also produced the spine-tingling roar that Lamborghini enthusiasts crave.

As the Gallardo evolved, so did its fuel injection system. The early models featured a highly advanced multi-point fuel injection system, which was already ahead of its time in terms of precision and efficiency. However, Lamborghini didn't rest on its laurels. In later models, such as the Gallardo LP560-4, direct fuel injection was introduced. This technological leap allowed for even more precise fuel delivery, injecting gasoline directly into the combustion chambers under extremely high pressure.

The result was a significant improvement in both fuel efficiency and power output, demonstrating Lamborghini's commitment to continual refinement and advancement. Overseeing all these complex systems was a sophisticated engine management system. The Lamborghini LIE (Lamborghini Iniezione Elettronica) system was the brain of the operation, constantly monitoring and adjusting various parameters to ensure optimal performance across all driving conditions. This electronic wizardry allowed the V10 to deliver its power smoothly and consistently, whether cruising on the highway or pushing the limits on a racetrack. It also played a crucial role in meeting increasingly stringent emissions standards without compromising the Gallardo's exhilarating performance.

The culmination of these design elements resulted in an engine that was more than the sum of its parts. It was a powerplant that could deliver jaw-dropping performance while also being civilized enough for daily use, a combination that was relatively rare in the supercar world at the time. The Gallardo's V10 represented a new era for Lamborghini, one where cutting-edge technology and traditional supercar thrills coexisted in perfect harmony. This engine not only defined the Gallardo but also set the stage for future Lamborghini models, cementing the company's position at the forefront of automotive engineering.

Section 3.3: Performance Benchmarks

The Lamborghini Gallardo's V10 engine represented a significant leap forward in performance and technology, setting new benchmarks in the supercar world. When compared to its predecessors, the Gallardo's powerplant showcased remarkable advancements. The V10 was more compact and efficient than the Diablo's V12, yet it nearly matched its power output. This achievement highlighted the rapid progress in engine technology and Lamborghini's commitment to innovation.

Performance metrics for the Gallardo were nothing short of impressive. The base model's 5.0-liter V10 propelled the car from 0 to 60 mph in just 4.2 seconds, a figure that would improve with subsequent iterations. The Gallardo Superleggera, with its enhanced V10, shaved this time down to a blistering 3.4 seconds, firmly establishing the model as a top-tier performer in the supercar realm. These figures weren't just numbers on paper; they translated to exhilarating real-world performance that left drivers and passengers alike breathless.

One of the most remarkable aspects of the Gallardo's V10 was its reliability. In an arena where high-strung engines often came with significant maintenance demands, the Gallardo's powerplant proved surprisingly robust. It demonstrated better reliability in daily use than many of its exotic counterparts, a factor that contributed significantly to the model's popularity and accessibility. This reliability didn't come at the cost of performance; instead, it added to the Gallardo's appeal as a supercar that could be enjoyed regularly, not just as a garage queen.

The V10's adaptability was another key strength, allowing Lamborghini to create a range of Gallardo variants to suit different tastes and performance requirements. From the base model to high-performance versions, the engine proved remarkably flexible. In the Gallardo LP570-4 Superleggera, for instance, the V10 was tuned to produce a staggering 570 horsepower, showcasing the engine's potential when pushed to its limits. This adaptability allowed Lamborghini to keep the Gallardo fresh and competitive throughout its production run, continuously offering new and exciting variants to enthuse both new and existing customers.

Despite its focus on performance, the Gallardo's V10 also made strides in efficiency. In the context of supercar standards, the engine managed to balance its power output with reasonable fuel consumption and emissions. Later models of the Gallardo, equipped with technologies like direct injection, met stringent Euro 5 emissions standards.

This achievement was particularly noteworthy, given the engine's high output and the vehicle's performance-oriented nature. The Gallardo's V10 engine set new benchmarks not just in raw performance, but in the overall package it offered. It combined blistering acceleration and top speed with daily drivability and reliability. The engine's character is responsive, high-revving, and aurally intoxicating, becoming a defining feature of the Gallardo, contributing significantly to its driver appeal and market success.

Moreover, the V10's performance benchmarks weren't limited to straight-line speed. The engine's power delivery, combined with the car's all-wheel-drive system and balanced chassis, resulted in exceptional handling and track performance. The Gallardo quickly established itself as a formidable presence on racetracks around the world, both in professional motorsport and in the hands of enthusiast owners.

In essence, the performance benchmarks set by the Gallardo's V10 engine redefined expectations for supercars in the early 21st century. It proved that extreme performance could coexist with reliability and relative efficiency, setting a new standard that competitors would strive to match. The engine's capabilities played a crucial role in establishing the Gallardo as one of the most successful and influential supercars of its era, cementing Lamborghini's position at the forefront of high-performance automotive engineering.

Section 3.4: Evolution of the V10

The Lamborghini Gallardo's V10 engine underwent significant evolution throughout the model's decade-long production run, reflecting the relentless pursuit of performance and technological advancement that defines the supercar industry.

Lamborghini Gallardo: A Decade of Domination

At its introduction in 2003, the Gallardo's 5.0-liter V10 was already a marvel of engineering, producing an impressive 493 horsepower. However, Lamborghini's engineers were far from content with this initial offering. Over the years, they continuously refined and enhanced the engine, pushing the boundaries of naturally aspirated performance.

One of the most significant milestones in the V10's evolution came with the introduction of the Gallardo LP560-4 in 2008. This model featured an engine displacement increase to 5.2 liters, accompanied by a range of internal improvements. The result was a substantial power boost to 552 horsepower, living up to the "560" in the model name (which referenced the metric horsepower figure).

The most crucial technological leap in the V10's development was the implementation of direct fuel injection. This advanced system, which sprays fuel directly into the combustion chamber rather than the intake ports, brought multiple benefits. It improved fuel atomization, allowing for more precise control over the combustion process. This resulted in a remarkable 20% increase in fuel efficiency, a significant achievement for a high-performance engine. Moreover, direct injection contributed to a cleaner burn, helping the Gallardo meet increasingly stringent emissions standards without sacrificing performance.

The quest for more power continued throughout the Gallardo's lifespan. Lamborghini's engineers employed various techniques to extract every possible horsepower from the V10. They optimized intake and exhaust systems, refined the engine management software, and in some cases, utilized lighter internal components. These efforts culminated in the final iterations of the Gallardo, such as the LP570-4 Superleggera and Squadra Corse, where the V10 produced a phenomenal 570 horsepower.

Advancements in materials science played a crucial role in the evolution of the V10. Later models incorporated more exotic materials to enhance performance and durability.

For instance, titanium connecting rods were introduced in high-performance variants, which reduced reciprocating mass and enabled higher engine speeds. Similarly, the use of more heat-resistant alloys in critical components improved the engine's reliability under extreme conditions.

As the V10's power output increased, so did the challenges of managing heat. Lamborghini's engineers responded with continual improvements to the cooling system. Enhanced oil coolers were added to maintain optimal operating temperatures, particularly in high-performance variants and for use on the track. The cooling system's efficiency was further improved through better airflow management, with redesigned intakes and vents directing more air to critical components.

Throughout its evolution, the Gallardo's V10 maintained its fundamental character as a high-revving, naturally aspirated engine. While many competitors turned to forced induction to boost power, Lamborghini remained committed to the linear power delivery and instantaneous response that only a naturally aspirated engine can provide. This decision helped preserve the Gallardo's unique driving experience and aural signature, which became integral to its appeal.

The evolution of the Gallardo's V10 engine is a testament to Lamborghini's engineering prowess and commitment to continuous improvement. From its origins as a collaborative project with Audi to its final form as one of the most potent naturally aspirated engines in the automotive world, the V10 remained at the heart of the Gallardo's identity. Its development over the years not only kept the Gallardo competitive in a rapidly advancing market but also laid the groundwork for future Lamborghini powerplants, ensuring that the legacy of this remarkable engine would continue well beyond the Gallardo's production run.

Section 3.5: The V10's Impact on Driving Dynamics

The Lamborghini Gallardo's V10 engine was more than just a powerplant; it was the beating heart that defined the car's entire character and driving experience. This section examines how the engine's unique characteristics influenced the Gallardo's performance on both the road and the track. At the core of the Gallardo's dynamic prowess was the V10's contribution to weight distribution.

The engine's compact design allowed engineers to position it optimally within the chassis, resulting in a near-perfect 42/58 front/rear weight distribution. This balance was crucial in giving the Gallardo its renowned handling characteristics, providing a level of agility that belied its powerful nature. The mid-engine layout, made possible by the V10's relatively compact dimensions, ensured that the car's mass was centralized, enhancing its responsiveness to driver inputs.

The relationship between the V10 engine and the Gallardo's all-wheel-drive system was symbiotic, each complementing the other to deliver an unparalleled driving experience. The engine's broad torque curve, with ample power available throughout the rev range, was ideally suited to the AWD layout. This pairing allowed for exceptional traction in various driving conditions, from launching off the line to powering out of corners on a racetrack. The AWD system's ability to variably distribute power between the front and rear axles meant that the V10's substantial output could be effectively harnessed, regardless of road conditions or driving style.

Perhaps most significantly, the V10 engine played a crucial role in shaping the Gallardo's unique driving character. Unlike many of its turbocharged competitors, the Gallardo's naturally aspirated V10 delivered power in a beautifully linear fashion. This predictable and progressive power delivery gave drivers a sense of connection and control that was often lacking in forced-induction engines. The engine's willingness to rev, combined with its immediate throttle response, created a driving experience that was both exhilarating and accessible. It allowed drivers to precisely modulate power output, whether they were navigating city streets or attacking apexes on a circuit.

On the racetrack, the V10's characteristics truly shone. Its high-revving nature and responsive power delivery made it a formidable tool for track driving. The engine's ability to quickly build and sustain high RPMs allowed drivers to exploit the full breadth of its power band, crucial for maintaining momentum through a series of corners.

Moreover, the natural aspiration meant consistent power delivery lap after lap, without the heat management issues often associated with turbocharged engines under sustained high-load conditions. Remarkably, despite its high-performance capabilities, the V10 engine also contributed significantly to the Gallardo's everyday usability aspect, which set it apart in the supercar world.

The engine's tractable nature at low speeds, coupled with its smooth power delivery, made the Gallardo surprisingly manageable in urban environments. Unlike some high-strung exotic powerplants, the V10 was content to cruise at low RPMs, making stop-and-go traffic less of a chore than one might expect in a supercar.

This versatility was a key factor in the Gallardo's success. The V10 allowed the car to seamlessly transition from a comfortable grand tourer to a track-ready performance machine at a moment's notice. It provided the thrilling acceleration and sound expected of a Lamborghini, yet remained composed and drivable in less demanding situations.

The engine's reliability further enhanced the Gallardo's appeal as a usable supercar. While delivering exceptional performance, the V10 proved more dependable in daily use than many of its exotic counterparts, contributing to lower maintenance costs and increased owner satisfaction.

In essence, the V10 engine was the linchpin of the Gallardo's driving dynamics. It provided the power and character expected of a Lamborghini while offering a level of accessibility and versatility previously unseen in the brand's models.

This engine didn't just propel the car; it defined its personality, setting a new standard for what a modern supercar could be: thrilling yet usable, powerful yet manageable. The V10's impact on the Gallardo's driving dynamics was so profound that it not only shaped the car's legacy but also influenced the direction of Lamborghini's future models, cementing its place as one of the most significant engines in the marque's history.

Section 3.6: Comparison with Competitors

The Lamborghini Gallardo's V10 engine was a defining feature that set it apart in the competitive landscape of supercars. To truly appreciate its significance, it's essential to compare it with the power plants of its contemporaries.

When pitted against Ferrari's offerings of the time, the Gallardo's V10 presented a stark contrast in philosophy. While Ferrari increasingly turned to forced induction for their V8 engines, Lamborghini remained steadfast in their commitment to natural aspiration. This decision resulted in a more linear power delivery and a more visceral, high-revving character that many enthusiasts prized. The Gallardo's engine responded with immediate throttle response and a spine-tingling wail that turbocharged engines struggled to match. However, Ferrari's approach often yielded higher peak torque figures, especially in later models, giving them an edge in specific performance metrics.

Comparing the Gallardo's V10 to Porsche's renowned flat-six engines reveals another interesting contrast. Porsche's engines, particularly those found in high-end 911 models, were marvels of engineering in their own right, known for their efficiency and relatively compact design.

Yet, the Gallardo's V10 offered a level of exotic appeal that the more conventional Porsche powerplants couldn't quite match. The Lamborghini engine's larger displacement and additional cylinders provided a different kind of driving experience, one characterized by a broader power band and a more dramatic auditory experience.

American supercars of the era, such as the Ford GT, typically relied on large-displacement V8 engines. These powerplants were known for their prodigious torque output and muscular character. In contrast, the Gallardo's V10 represented a more high-strung, European approach to performance. It traded some low-end grunt for a thrilling high-rev experience, encouraging drivers to explore the upper reaches of the tachometer.

This difference in character meant that while American V8s might have had an advantage in straight-line acceleration, especially from a standing start, the Gallardo's engine often felt more at home on winding roads or race tracks where its rev-happy nature could shine.

Interestingly, the Gallardo's V10 shared some DNA with other engines in the Volkswagen Audi Group (VAG) portfolio, particularly those found in Audi's RS models. However, it would be a mistake to consider them identical. The Lamborghini engineers tuned and modified the engine significantly, optimizing it for supercar performance. The result was an engine that, while sharing some basic architecture with its Audi cousins, delivered a distinctly more exotic character and considerably more power.

Perhaps the most intriguing comparison is with Lamborghini's own V12 engines, such as those found in the contemporary Murciélago. The V10 and V12 represented two different approaches to supercar performance within the same brand. The V12 was all about brute force and top-end power, a continuation of Lamborghini's traditional flagship approach.

The Gallardo's V10, on the other hand, offered a more balanced package. It was lighter, more compact, and more responsive, allowing for a different kind of driving experience. While it couldn't match the outright power of the V12, it provided a more accessible performance envelope that many drivers found more usable in real-world conditions.

In the context of these comparisons, the Gallardo's V10 emerges as a unique and compelling offering. It combined elements of high-revving exotic performance with a degree of accessibility unusual in the supercar world. This engine played a crucial role in establishing the Gallardo as a success, proving that Lamborghini could compete in a broader market segment without sacrificing the brand's core values of performance and excitement. The V10's ability to hold its own against a diverse array of competitors underscores its significance not just to Lamborghini but to the evolution of supercar engines as a whole.

Section 3.7: Legacy of the Gallardo's V10

The Lamborghini Gallardo's V10 engine left an indelible mark on the automotive world, shaping not only the future of Lamborghini but also influencing the broader supercar landscape. This powerplant's legacy extends far beyond its decade-long tenure in the Gallardo, continuing to resonate in the halls of Sant'Agata Bolognese and beyond.

The engine's influence on future Lamborghini models cannot be overstated. When the time came to develop the Gallardo's successor, the Huracán, Lamborghini engineers didn't start from scratch. Instead, they evolved the Gallardo's V10, refining and enhancing its already formidable capabilities.

The Huracán's powerplant, while thoroughly retuned, carries the DNA of its predecessor, maintaining the high-revving, naturally aspirated chassis that became synonymous with Lamborghini's V10. This continuity speaks volumes about the original engine's design and potential, proving that its fundamental architecture was sound enough to form the basis for the next generation of Lamborghini supercars.

The Gallardo's V10 also had a profound impact on the broader supercar market. Its success prompted other manufacturers to reconsider their engine strategies. While V8 and V12 engines had long been the staples of the supercar world, the Gallardo demonstrated that a V10 could offer a compelling alternative, blending the high-revving nature of smaller engines with the power output traditionally associated with larger ones. This influence can be seen in competitors' subsequent models and engine development programs, as they sought to capture some of the magic that made the Gallardo's powerplant so special.

Lamborghini Gallardo: A Decade of Domination

One of the most significant aspects of the V10's legacy is its contribution to Lamborghini's brand image. Before the Gallardo, Lamborghini was often viewed as a producer of exotic, but somewhat temperamental, supercars.

The V10 engine, with its balance of performance and relative accessibility, helped broaden Lamborghini's appeal. It maintained the brand's reputation for producing thrilling, high-performance vehicles while introducing a level of usability that attracted a wider range of enthusiasts. This shift in perception played a crucial role in Lamborghini's growth and success in the 21st century, helping to position the brand as a more rounded and capable supercar manufacturer.

The engine's legacy extends to the racetrack as well. Its robust design and impressive power output made it an excellent platform for motorsport applications. The Gallardo, powered by variations of this V10, found success in various GT3 racing series around the world. This racing pedigree not only enhanced the engine's reputation but also provided valuable data and experience that fed back into Lamborghini's road car development programs.

In the annals of automotive history, the Gallardo's V10 holds a special place. As the industry moves increasingly towards downsized, turbocharged engines and electrification, the Gallardo's naturally aspirated V10 is often cited as one of the last significant engines of its kind. Its combination of high-revving excitement, linear power delivery, and evocative sound has made it a favorite among enthusiasts and a benchmark against which modern engines are still measured.

The engine's influence can even be seen in Lamborghini's approach to future technologies. As the company develops hybrid and electric powertrains, the lessons learned from the V10, particularly in terms of power delivery and driving experience, continue to inform its engineering decisions. Lamborghini's commitment to maintaining the emotional connection between driver and machine, so perfectly exemplified by the Gallardo's V10, remains a guiding principle as they navigate the challenges of a changing automotive landscape.

In conclusion, the legacy of the Gallardo's V10 engine is multi-faceted and far-reaching. It reshaped Lamborghini's image, influenced the design of competitors, proved its mettle on racetracks, and set a benchmark for supercar engines that remains in place today. More than just a powerplant, it became the beating heart of a car that transformed Lamborghini, leaving an enduring impact on the automotive world and securing its place in the pantheon of legendary engines.

Lamborghini Gallardo: A Decade of Domination

Chapter 4: Evolution of Power: Engine Developments Through the Years

Section 4.1: The Original 5.0-liter V10

The Lamborghini Gallardo made its debut in 2003, introducing the world to a new era of exotic performance from the iconic Italian manufacturer. At the heart of this revolutionary supercar lay a powerplant that would set the tone for a decade of automotive excellence: the original 5.0-liter V10 engine.

This engineering marvel was a testament to Lamborghini's commitment to pushing the boundaries of performance and design. The engine boasted impressive specifications, producing a formidable 493 horsepower at a screaming 7,800 rpm, complemented by 376 lb-ft of torque peaking at 4,500 rpm. These numbers weren't just impressive on paper; they translated into blistering real-world performance that would cement the Gallardo's place in supercar history.

Lamborghini Gallardo: A Decade of Domination

Lamborghini's decision to equip the Gallardo with a V10 engine was a strategic one, carefully considered to position the new model within the company's lineup. While the flagship models like the Murciélago carried on the tradition of V12 power, the V10 configuration allowed the Gallardo to carve out its own unique identity. This choice struck a perfect balance between the exotic appeal expected of a Lamborghini and a more accessible performance package that would broaden the brand's appeal.

The V10's character was defined by its high-revving nature and linear power delivery. Unlike some turbocharged competitors, the Gallardo's naturally aspirated engine provided an immediate and predictable response to throttle inputs. This characteristic became a hallmark of the Gallardo driving experience, offering drivers a visceral connection to the machine that was both thrilling and intuitive.

In crafting this powerplant, Lamborghini's engineers prioritized both performance and weight management. The engine block and cylinder heads were constructed from aluminum alloy, a choice that significantly reduced the overall mass of the powertrain. This attention to weight distribution was crucial in achieving the balanced handling dynamics that would become another defining feature of the Gallardo.

When compared to its contemporaries, the Gallardo's V10 stood out as a unique proposition in the segment. While Ferrari's F430 relied on a V8 and Porsche's 911 Turbo utilized a flat-six, the Gallardo's V10 offered a distinctive blend of power, sound, and character. It provided the exotic flair expected of a Lamborghini while delivering performance that could go toe-to-toe with its rivals.

The engine's soundtrack was another area where the V10 configuration proved its worth. The unique firing order and exhaust tuning resulted in a spine-tingling howl that became instantly recognizable. This aural signature added an emotional dimension to the Gallardo's appeal, ensuring that it not only performed like a true Lamborghini but also sounded like one.

As the foundation for future developments, the original 5.0-liter V10 was designed with evolution in mind. Its robust architecture would allow for subsequent increases in displacement and power output, ensuring that the Gallardo could remain competitive throughout its production run.

In essence, the original 5.0-liter V10 was more than just an engine; it was the beating heart of a car that would redefine Lamborghini for a new generation. It embodied the perfect blend of performance, character, and engineering prowess that would set the stage for a decade of refinement and improvement, cementing the Gallardo's place as one of the most significant supercars of its era.

Section 4.2: The 5.2-liter V10 Upgrade

In 2008, Lamborghini took a significant step in the evolution of the Gallardo's powerplant by increasing its displacement to 5.2 liters. This move was driven by the company's relentless pursuit of performance and its desire to keep the Gallardo competitive in an increasingly fierce supercar market.

The decision to upgrade the engine wasn't taken lightly. Lamborghini's engineers faced the challenge of extracting more power and torque from the V10 while maintaining the reliability and drivability that had become hallmarks of the Gallardo. The increase in displacement from 5.0 to 5.2 liters required a comprehensive redesign of several key engine components.

To achieve the larger displacement, both the bore and stroke of the engine were increased. This necessitated a redesign of the cylinder heads and pistons to accommodate the new dimensions. The engineering team also took this opportunity to optimize the combustion chambers and valve timing, further enhancing the engine's efficiency and power output.

The results of this upgrade were impressive. The new 5.2-liter V10, first introduced in the Gallardo LP560-4, produced a formidable 552 horsepower - a significant jump from the original engine's 493 hp. This power increase wasn't just about bragging rights; it transformed the Gallardo's performance envelope, improving acceleration times and enhancing the car's already formidable capabilities on both road and track.

The additional displacement also resulted in increased torque, improving the engine's flexibility and responsiveness across the rev range. This was particularly noticeable in everyday driving situations, where the extra low-end torque made the Gallardo even more tractable and enjoyable to drive at moderate speeds.

However, the implementation of the new engine wasn't without its challenges. The increased power output meant that the engineering team had to balance performance gains with thermal management and reliability considerations carefully. The system was upgraded to handle the additional heat generated by the more powerful engine, and internal components were strengthened to ensure longevity under the increased stresses.

The 5.2-liter upgrade also had a profound impact on the Gallardo's character. While the original 5.0-liter engine was already renowned for its high-revving nature and distinctive sound, the larger displacement V10 added a new dimension to the auditory experience. The engine note became even more aggressive, with a deeper growl at low revs and an even more intoxicating howl at the upper reaches of the tachometer.

From a competitive standpoint, the 5.2-liter V10 allowed the Gallardo to assert its dominance in the segment. While rivals like the Ferrari F430 were still using V8 engines, Lamborghini's V10 offered a unique proposition, balancing appeal with raw performance. The larger engine helped to differentiate the Gallardo from its competitors further, reinforcing its position as one of the most desirable supercars in its class.

The introduction of the 5.2-liter V10 wasn't just a one-time upgrade; it set the stage for further developments throughout the remainder of the Gallardo's production run. This engine would serve as the basis for even more powerful variants in the future, demonstrating the inherent potential of its design.

In essence, the upgrade to the 5.2-liter V10 represented a pivotal moment in the Gallardo's evolution. It showcased Lamborghini's commitment to continuous improvement and its ability to extract even more performance from an already impressive powerplant. This engine would go on to define the Gallardo's character in its later years, cementing its legacy as one of the significant supercar engines of its era.

Section 4.3: Direct Injection Implementation

The introduction of direct fuel injection marked a significant milestone in the evolution of the Lamborghini Gallardo's V10 engine. This advanced technology revolutionized the way fuel was delivered to the engine, offering a host of benefits that enhanced both performance and efficiency.

Direct injection is a fuel delivery system that introduces fuel directly into the combustion chamber of each cylinder, rather than mixing it with air in the intake manifold. This precise method of fuel delivery allows for more accurate control over the fuel-air mixture, resulting in improved combustion efficiency, increased power output, and reduced fuel consumption.

Lamborghini first implemented direct injection in the 2009 Gallardo LP560-4, marking a first for the company and showcasing their commitment to cutting-edge engine technology. The integration of this system required a significant redesign of the engine's cylinder heads and the development of new fuel system components. High-pressure fuel pumps and specially designed injectors were incorporated to deliver fuel at the extremely high pressures required for direct injection to function effectively.

The results of this technological advancement were immediately apparent. The direct-injected 5.2-liter V10 produced an impressive 560 horsepower, a notable increase from its predecessor. This power boost was achieved while simultaneously improving fuel efficiency by approximately 18%, demonstrating that high performance and improved economy were not mutually exclusive.

The implementation of direct injection also allowed for a higher compression ratio, as the cooling effect of the fuel being injected directly into the cylinder helped prevent pre-ignition and knock. This higher compression ratio further contributed to the engine's increased efficiency and power output.

Moreover, the direct injection system provided the Gallardo's engine with improved throttle response. The precise control over fuel delivery allowed for near-instantaneous adjustments to engine output, resulting in a more responsive and engaging driving experience.

However, the integration of direct injection was not without its challenges. Lamborghini's engineers had to overcome several hurdles in the development process. One significant challenge was managing the increased heat generated by the direct injection system. The high-pressure fuel pump and injectors operate at extreme pressures, generating additional heat that needs to be effectively managed to ensure long-term reliability.

Another challenge was the potential for carbon buildup on intake valves, a common issue with direct injection engines. In port injection systems, the fuel spray helps keep the intake valves clean; however, with direct injection, the fuel no longer passes over these valves. Lamborghini had to develop strategies to mitigate this issue, including advanced engine management algorithms and specific maintenance procedures.

When compared to its competitors, Lamborghini's implementation of direct injection in the Gallardo was notably advanced. While some rivals, such as Ferrari, had already introduced direct injection in models like the F430 Scuderia, Lamborghini's system was uniquely tailored to the V10 configuration. This bespoke approach allowed Lamborghini to maximize the benefits of direct injection while maintaining the distinctive character of the Gallardo's engine.

The introduction of direct injection in the Gallardo not only improved its performance metrics but also helped the model comply with increasingly stringent emissions regulations. The more efficient combustion process resulted in lower emissions, allowing the Gallardo to meet stricter environmental standards without compromising its thrilling performance.

In conclusion, the implementation of direct injection in the Lamborghini Gallardo's V10 engine represented a significant leap forward in the model's evolution. It exemplified Lamborghini's commitment to embracing advanced technologies to enhance performance, efficiency, and environmental compatibility. This advancement set the stage for future developments in Lamborghini's engine technology, influencing not only subsequent iterations of the Gallardo but also paving the way for the next generation of Lamborghini supercars.

Section 4.4: High-Performance Variants

Throughout its illustrious production run, Lamborghini consistently pushed the boundaries of performance with the Gallardo, offering a series of high-performance variants that showcased the true potential of its remarkable V10 engine. These special editions not only delivered increased power and performance but also served as testbeds for new technologies and engineering solutions that would eventually benefit the entire Gallardo lineup.

The evolution of these high-performance variants began with the introduction of the Gallardo Superleggera in 2007. This lightweight edition saw the original 5.0-liter V10 engine tuned to produce 523 horsepower, a notable increase from the standard model. The power boost was achieved through meticulous engine tuning and the implementation of a revised exhaust system. The Superleggera's engine modifications weren't just about raw power; they also enhanced throttle response and overall engine dynamics, providing a more engaging driving experience.

As the Gallardo platform matured, so did the ambition of Lamborghini's engineers. The introduction of the Gallardo Performante in 2011 marked another significant milestone in the model's engine development. Based on the Spyder variant, the Performante featured a retuned version of the 5.2-liter V10, now producing a formidable 570 horsepower. This increase in power was accompanied by improvements in torque delivery and engine response, further emphasizing the Performante's track-focused nature.

The Super Trofeo Stradale edition, introduced in 2011, represented a bold step in bringing racetrack technology to the street. This limited-edition model borrowed heavily from the Gallardo Super Trofeo race car, featuring a highly tuned version of the 5.2-liter V10 that produced 570 horsepower.

The engine in the Super Trofeo Stradale wasn't just about peak power; it was engineered to deliver exceptional performance across the entire rev range, making it equally adept on both road and track.

The culmination of the Gallardo's engine development came with the LP 570-4 Squadra Corse, one of the final and most extreme variants of the model. Named after Lamborghini's racing division, this edition boasted 570 horsepower and represented the pinnacle of the Gallardo's V10 engine evolution. The Squadra Corse's engine was a masterpiece of engineering, incorporating all the lessons learned from previous high-performance variants and the company's racing experience.

What made these high-performance variants truly special was not just their increased power output but the holistic approach Lamborghini took in their development. Each model featured carefully calibrated engine management systems, optimized to extract maximum performance while ensuring reliability. The exhaust systems were often bespoke, designed not only to reduce back pressure and increase power but also to produce an even more intoxicating soundtrack.

Moreover, these variants served as rolling laboratories, allowing Lamborghini to test and refine technologies that would eventually trickle down to other models. Advancements in areas such as thermal management, friction reduction, and materials technology were often first implemented in these high-performance editions before being adapted for broader use.

The creation of these high-performance variants also highlighted Lamborghini's responsiveness to market demands and its commitment to pushing the envelope of what was possible with the Gallardo platform. Each new variant not only showcased the company's engineering prowess but also reignited interest in the model, helping to maintain its appeal throughout its decade-long production run.

In essence, the high-performance variants of the Gallardo tell a story of relentless pursuit of excellence. From the original Superleggera to the final Squadra Corse, each model represented a significant step forward in the evolution of the Gallardo's V10 engine. These special editions not only delivered exceptional performance but also played a crucial role in the continuous development and refinement of one of the most successful engines in Lamborghini's history.

Section 4.5: Advancements in Engine Management Systems

As the Lamborghini Gallardo evolved over its decade-long production run, so did the sophisticated electronic brain controlling its powerful V10 heart. The engine management system played a crucial role in extracting maximum performance from the Gallardo's powerplant while ensuring reliability and compliance with ever-tightening emissions regulations.

In the early years of the Gallardo, the engine management system was already advanced for its time, but it was just the beginning of a journey towards even greater sophistication. As computing power increased and sensor technology improved, Lamborghini's engineers were able to implement more complex and precise control strategies.

One of the most significant areas of improvement was in fuel mapping. The original Gallardo's fuel maps were well-tuned, but as the engine evolved, so did the complexity and precision of these maps. Later models featured incredibly detailed three-dimensional fuel maps that could adjust fuel delivery based on a multitude of parameters such as engine speed, load, temperature, and even altitude. This allowed for optimized power delivery across the entire rev range while simultaneously improving fuel efficiency and reducing emissions.

Ignition timing control saw similar advancements throughout the Gallardo's lifespan. Early models already featured variable timing, but the level of control became increasingly precise in later iterations. Enhanced ignition timing control allowed engineers to push the envelope further, implementing more aggressive timing strategies without risking engine damage. This resulted in improved power output and throttle response, particularly in the high-rev ranges where the Gallardo's V10 truly sang.

The integration of advanced sensors played a pivotal role in these improvements. As sensor technology progressed, the engine management system gained access to more accurate and real-time data about the engine's operating conditions. Particularly noteworthy was the introduction of more precise knock sensors. These allowed the engine to operate closer to its theoretical limits by detecting the onset of detonation earlier and with greater accuracy. The system could then make split-second adjustments to prevent damage while maintaining maximum performance.

Another area of significant advancement was the integration between the engine management system and the vehicle's traction and stability control systems. In the early Gallardo models, these systems operated independently. However, as the model evolved, the integration became more seamless. Later Gallardo variants featured engine management systems that worked in close concert with the stability control, adjusting power delivery in milliseconds based on traction conditions. This not only enhanced performance but also improved safety, allowing drivers to explore the car's limits with greater confidence.

The introduction of driver-selectable modes in later Gallardo models further showcased the advancements in engine management. These modes could alter various parameters of the engine's behavior, from throttle response to redline, allowing drivers to tailor the car's character to different driving scenarios. Whether it was a relaxed cruise or a hot lap on a racetrack, the engine management system could adapt to deliver the optimal driving experience.

Emissions control was another area where the evolving engine management system played a crucial role. As regulations became stricter, particularly in Europe, Lamborghini had to ensure the Gallardo remained compliant without sacrificing its thrilling performance. Advanced catalytic converter systems, working in tandem with more precise fueling and ignition control, reduced harmful emissions while maintaining the Gallardo's characteristic power and sound.

The culmination of these advancements was most evident in the final iterations of the Gallardo, such as the LP 570-4 Squadra Corse. These models showcased engine management systems that were light-years ahead of the original Gallardo in terms of complexity and capability. They allowed the V10 engine to produce prodigious power while meeting stringent emissions standards, a balance that would have been impossible with earlier technology.

In conclusion, the evolution of the Gallardo's engine management system is a testament to the rapid pace of automotive technology during the car's lifetime. From more precise control of fundamental parameters to complex integration with other vehicle systems, these advancements played a crucial, if often unseen, role in maintaining the Gallardo's position as one of the most thrilling and capable supercars of its era. The lessons learned and technologies developed during this period laid the groundwork for the even more advanced systems found in Lamborghini's current models, ensuring that the legacy of the Gallardo lives on in the digital brains of its successors.

Section 4.6: Exhaust System Developments

The exhaust system of a supercar is far more than just a means to expel combustion gases; it's an integral part of the engine's performance and a key component in crafting the vehicle's aural signature. Throughout the Lamborghini Gallardo's production run, the exhaust system underwent significant developments, each iteration enhancing both performance and the emotive sound that is so crucial to the supercar experience.

When the Gallardo first roared onto the scene in 2003, its exhaust system was already a carefully tuned piece of engineering. Lamborghini's engineers had to strike a delicate balance between optimizing back pressure for performance, meeting increasingly stringent noise regulations, and delivering the spine-tingling sound expected of a Lamborghini. The original system featured stainless steel construction and a relatively straightforward design, with primary focus on flow characteristics to complement the 5.0-liter V10's power delivery.

As the Gallardo evolved, so did its exhaust system. One of the most significant advancements was in material technology. Later models incorporated more exotic materials, with titanium becoming a popular choice for high-performance variants. Titanium offered several advantages: it's significantly lighter than stainless steel, helping to reduce overall vehicle weight; it has superior heat management properties, allowing the exhaust to operate more efficiently at high temperatures; and it possesses unique acoustic properties that contribute to an even more exhilarating exhaust note.

The introduction of variable exhaust systems marked another leap forward in the Gallardo's exhaust development. This technology, featuring electronically controlled valves, allowed the exhaust note to be altered on demand. In normal driving conditions, the valves would remain closed, routing exhaust gases through sound-dampening chambers to maintain civilized noise levels.

However, at the push of a button or under heavy acceleration, these valves would open, allowing exhaust gases to bypass the quieter route and flow through a more direct, less restrictive path. This not only produced a more aggressive and intoxicating sound but also reduced back pressure, liberating additional horsepower at high RPMs.

The impact of these exhaust developments on the Gallardo's performance and character cannot be overstated. In terms of performance, the more advanced systems contributed to incremental power gains throughout the model's lifespan. The reduced back pressure and more efficient scavenging of exhaust gases allowed the engine to breathe more freely, especially at high RPMs where every horsepower counts.

Perhaps even more significant was the impact on the Gallardo's aural character. The sound of a Lamborghini has always been a crucial part of its appeal, and the later iterations of the Gallardo took this to new heights. The combination of advanced materials, variable valve systems, and meticulous tuning yielded an exhaust note that was both more aggressive and more refined. At low RPMs, the engine would purr with a deep, muscular tone. But as the revs climbed, it would build to a spine-tingling crescendo, a high-pitched, almost operatic wail that became one of the Gallardo's most distinctive and celebrated features.

Lamborghini's continuous development of the Gallardo's exhaust system demonstrates the company's commitment to enhancing every aspect of the supercar experience. From the early models to the final editions, each iteration brought improvements in performance, weight reduction, and acoustic engineering. The result was not just a more capable car, but one that engaged the driver's senses more completely, adding an extra layer of emotion to the already exhilarating experience of piloting a Lamborghini. The evolution of the Gallardo's exhaust system stands as a testament to the fact that in the world of supercars, even the most minor details can have a profound impact on the overall experience.

Section 4.7: Reliability and Durability Improvements

In the realm of high-performance supercars, reliability is often as crucial as raw power. As the Lamborghini Gallardo's engine evolved over its production run, becoming more powerful and sophisticated, the engineers at Sant'Agata Bolognese faced the challenge of maintaining and even improving reliability under increasingly demanding conditions. This section explores the various improvements made to ensure the Gallardo's V10 engine remained as dependable as it was potent.

The importance of reliability in high-performance engines cannot be overstated. While supercars are often celebrated for their extreme power outputs and blistering acceleration times, the ability to deliver this performance consistently and over extended periods is what separates truly great engines from merely good ones. As the Gallardo's engine evolved from its original 5.0-liter configuration to the final 5.2-liter form, Lamborghini had to ensure that each iteration not only produced more power but also maintained the robustness expected of a premium supercar.

One of the primary areas of focus for improving reliability was the cooling system. As the Gallardo's engine became more powerful over time, it naturally produced more heat, necessitating enhancements to the cooling capabilities. Later Gallardo models featured redesigned cooling systems with improved radiators, enhanced coolant flow, and more efficient heat exchangers. These improvements allowed the engine to maintain optimal operating temperatures even under extreme conditions, such as high-speed track driving or in hot climates. For instance, the Gallardo LP 570-4 Superleggera features additional cooling vents and a more efficient coolant pump, which contribute to its ability to sustain high performance for extended periods.

Advancements in lubrication systems played a crucial role in enhancing the engine's longevity. Lamborghini introduced more advanced synthetic oils specifically formulated for the high-stress conditions experienced by the V10 engine. These oils provided better protection against wear and improved heat dissipation. Additionally, oil pump designs were refined to ensure consistent oil pressure and flow, even during high-G cornering and braking. The introduction of oil coolers in later models further helped maintain optimal oil temperatures, crucial for protecting engine internals during high-performance driving.

As the power output of the Gallardo's engine increased, so did the stress on internal components. To address this, Lamborghini continuously upgraded various engine parts for increased durability. Later Gallardo engines featured strengthened connecting rods capable of withstanding higher forces, more robust valve train components to handle increased rpm, and pistons made from advanced alloys for improved heat resistance. The cylinder heads were also reinforced to maintain proper sealing under higher combustion pressures. These upgrades allowed the engine to produce more power while maintaining its high-revving character reliably.

The implementation of more sophisticated engine management systems also contributed significantly to the Gallardo's improved reliability. Advanced sensors and more powerful ECUs allowed for more precise control over fuel injection, ignition timing, and other critical parameters. This not only optimized performance but also helped prevent conditions that could lead to engine damage, such as detonation or over-revving. The ability to more finely tune the engine's operation for different conditions - from daily driving to track use - ensured that the engine was continually operating within safe parameters.

Long-term reliability data from Gallardo owners and service centers provide compelling evidence of these improvements. While early models were already known for being more reliable than many of their exotic counterparts, later Gallardos exhibited even better long-term durability. Many high-mileage examples, particularly from the latter half of the production run, have demonstrated remarkable resilience, with some surpassing 100,000 miles while still performing admirably. This real-world data stands as a testament to Lamborghini's continuous engineering refinements throughout the Gallardo's lifespan.

The focus on reliability and durability improvements in the Gallardo's engine was not just about preventing failures; it was about building confidence. Owners of later model Gallardos often report a sense of assurance in their car's ability to perform consistently, whether on a cross-country grand tour or during an intense track day. This reliability has contributed significantly to the Gallardo's reputation and desirability in the used supercar market.

In conclusion, the evolution of the Lamborghini Gallardo's V10 engine is a story not just of increasing power and performance, but also of enhancing reliability and durability. Through continuous improvements in cooling, lubrication, internal components, and engine management, Lamborghini ensured that the Gallardo's heart remained as robust as it was powerful. This commitment to reliability has played a crucial role in cementing the Gallardo's status as one of the most successful and beloved supercars of its era, capable of delivering thrilling performance time and time again.

Chapter 5: Shifting Gears: Transmission Innovations in the Gallardo

Section 5.1: The E-Gear: Lamborghini's Automated Manual Transmission

The Lamborghini Gallardo's introduction in 2003 marked a significant milestone in the company's history, not just for its sleek design and powerful engine, but also for its revolutionary transmission system. The E-Gear, Lamborghini's automated manual transmission, represented a bold step into the future of supercar technology.

The E-Gear system was Lamborghini's answer to the growing demand for faster, more efficient gear changes in high-performance vehicles. It combined the raw mechanical feel of a traditional manual gearbox with the speed and precision of computer-controlled shifting. This innovative system allowed drivers to change gears via paddle shifters mounted behind the steering wheel, eliminating the need for a clutch pedal and manual gear lever.

At its core, the E-Gear was a sophisticated piece of engineering. It utilized hydraulic actuators controlled by an electronic control unit to change gears in as little as 150 milliseconds. This lightning-fast shift time was a game-changer in the supercar world, allowing the Gallardo to accelerate more quickly and smoothly than ever before. The system's ability to match engine revs perfectly during downshifts also enhanced the car's stability during high-speed cornering and braking.

As the Gallardo evolved, so did the E-Gear system. Lamborghini's engineers continuously refined and improved the technology, pushing the boundaries of what was possible. By 2008, with the introduction of the Gallardo LP560-4, the E-Gear had been upgraded to reduce shift times to an astonishing 120 milliseconds. This improvement not only enhanced the car's performance but also provided an even more seamless and exhilarating driving experience.

The E-Gear system fundamentally changed the way drivers interacted with the Gallardo. It allowed for both hands to remain on the steering wheel at all times, enhancing control during high-speed maneuvers and on the racetrack. The system also offered multiple driving modes, from a more relaxed automatic setting for everyday driving to an aggressive sport mode that held gears longer and executed lightning-fast shifts for maximum performance.

However, like any groundbreaking technology, the E-Gear system was not without its critics. Some purists argued that it removed some of the raw, mechanical feel that made earlier Lamborghinis so engaging to drive. They claimed that the automated system took away the skill and satisfaction of executing a perfect manual gear change. Additionally, early versions of the E-Gear were occasionally criticized for their jerky operation at low speeds, though this was addressed mainly in later iterations.

Despite these criticisms, the E-Gear system played a crucial role in shaping the Gallardo's identity and success. It bridged the gap between traditional manual transmissions and the fully automatic gearboxes that would later become prevalent in the supercar world. The system's ability to offer both lightning-fast shifts for track use and smooth, automatic operation for daily driving made the Gallardo a more versatile and accessible supercar.

The E-Gear also paved the way for future transmission innovations in Lamborghini's lineup. The lessons learned and technologies developed for the Gallardo's E-Gear would go on to influence the design of transmission systems in subsequent models, including the Huracán and Aventador.

In the grand narrative of the Lamborghini Gallardo, the E-Gear stands out as a symbol of the car's forward-thinking design and Lamborghini's commitment to pushing the boundaries of automotive technology. It represented a crucial step in the evolution of supercar transmissions, setting new standards for performance and driver engagement that would influence the industry for years to come.

Section 5.2: The Six-Speed Manual: A Nod to Tradition

While the E-Gear system represented a leap forward in transmission technology, Lamborghini recognized the enduring appeal of a traditional manual gearbox. The six-speed manual transmission offered in the Gallardo was more than just an alternative; it was a testament to the company's commitment to delivering a pure, engaging driving experience.

The importance of the manual option in the Gallardo cannot be overstated. In an era where many supercar manufacturers were moving exclusively to automated transmissions, Lamborghini chose to cater to enthusiasts who preferred a more traditional, hands-on approach to driving. This decision reflected a deep understanding of their customer base, many of whom valued the visceral connection between driver and machine that only a manual transmission could provide.

The six-speed manual transmission in the Gallardo was a marvel of engineering in its own right. It featured a gated shifter, a hallmark of classic Italian supercars that added both visual appeal and tactile satisfaction to the driving experience. The precise metal-on-metal click as the driver moved through the gears became a signature sound of the manual Gallardo, evoking memories of legendary Lamborghinis of the past.

Throughout the Gallardo's production run, the manual transmission saw its own evolution and refinement. Later models benefited from improved shift feel and more precise gear engagement. These enhancements were the result of continuous feedback from drivers and Lamborghini's commitment to perfecting every aspect of the driving experience. The clutch action was also fine-tuned over the years, striking a balance between ease of use in everyday driving and the firm, positive engagement needed for high-performance situations.

When comparing the performance of manual and E-Gear-equipped Gallardos, the results were often surprising. While E-Gear cars typically posted faster acceleration times due to their ability to change gears more quickly and consistently, manual Gallardos were widely praised for their more involving driving experience. The ability to perfectly time a shift, to feel the car's balance change as the clutch engaged, and to execute a perfect heel-and-toe downshift were experiences that many drivers felt were worth the trade-off in outright performance.

However, as the Gallardo's production run progressed, the manual transmission option faced a declining popularity that mirrored broader industry trends. By 2013, fewer than 5% of Gallardo buyers opted for the manual transmission, signaling a significant shift in supercar buyer preferences. This trend was driven by several factors, including the improved performance and ease of use of automated systems, as well as shifts in demographics within the supercar market.

Despite its waning popularity, the manual Gallardo remained a highly sought-after model among enthusiasts and collectors. The rarity of manual examples, particularly in later production years, has led to these versions often commanding a premium in the used market. For many, the manual Gallardo represents the last of a breed, a supercar that combines breathtaking performance with the intimate involvement of a traditional gearbox.

The story of the manual transmission in the Gallardo is one of preserving tradition in the face of technological advancement. It stands as a testament to Lamborghini's understanding that for some drivers, the joy of driving isn't just about outright speed or lap times, but about the experience and the connection between car and driver. In offering and maintaining the manual option throughout the Gallardo's lifespan, Lamborghini ensured that this iconic model would cater to all types of enthusiasts, from those embracing the latest technology to those seeking a more classic supercar experience.

Section 5.3: All-Wheel Drive vs. Rear-Wheel Drive Transmissions

The Lamborghini Gallardo, which revolutionized the drivetrain configuration, is a testament to the brand's commitment to innovation and the driver's experience. When the Gallardo first roared onto the supercar scene in 2003, it introduced Lamborghini's first production all-wheel drive system, a technological marvel that set it apart from its predecessors and many of its competitors.

The all-wheel drive system in the Gallardo was a technological tour de force, derived from Audi's renowned Quattro technology. This implementation required a unique transmission setup to distribute power to all four wheels effectively. The system was designed to provide optimal traction and handling in various driving conditions, a significant departure from the rear-wheel drive layout traditionally associated with Lamborghini's V12 flagships.

At the heart of the Gallardo's AWD system was a viscous coupling that could continuously adjust the power distribution between the front and rear axles. In normal driving conditions, the system typically sends 70% of the engine's power to the rear wheels, maintaining a rear-biased feel that supercar enthusiasts craved. However, the true magic of this system lies in its ability to instantly adjust this power split based on driving conditions and driver input. In situations where additional traction was needed, such as during hard acceleration or in low-grip scenarios, the system could send up to 50% of the power to the front wheels, ensuring maximum traction and stability.

This AWD system integrated seamlessly with the Gallardo's transmission, whether it was the six-speed manual or the E-Gear automated manual. The result was a car that could put its considerable power down to the road with remarkable efficiency, leading to blistering acceleration times and confidence-inspiring handling in a variety of conditions.

However, as the Gallardo evolved, Lamborghini recognized the desire among some purists for a more traditional, rear-wheel drive experience. In 2009, they introduced the Gallardo LP550-2, marking a return to rear-wheel drive in a V10 Lamborghini. This variant featured a simplified transmission layout, with power being sent exclusively to the rear wheels.

The introduction of rear-wheel drive models alongside the AWD versions created an interesting dynamic in the Gallardo lineup. While the AWD models offered superior traction, especially in adverse conditions, and could generally put down faster acceleration times, the RWD models were often praised for their more engaging and playful handling characteristics.

The rear-wheel drive layout reduced weight and complexity, and many driving enthusiasts appreciated the more traditional supercar feel it provided, with its greater potential for controllable oversteer and a more direct connection between driver inputs and vehicle responses.

The performance and handling differences between AWD and RWD Gallardos were notable. AWD models excelled in providing sure-footed traction, allowing drivers to confidently exploit the car's performance potential in a wider range of conditions. They were particularly advantageous in wet or low-grip situations, where the ability to distribute power to all four wheels provided a significant traction advantage.

On the other hand, RWD models, while potentially more challenging to drive at the limit, offered a rawer, more traditional supercar experience. They were often described as more engaging and rewarding for skilled drivers, providing a purer form of driver-car interaction.

The development and refinement of both AWD and RWD variants allowed Lamborghini to advance their transmission technologies significantly. Engineers had to optimize gear ratios, differential settings, and electronic control systems for both drivetrain layouts, leading to advancements that benefited the entire Gallardo range. This dual-pronged approach also provided valuable insights that would influence future Lamborghini models, helping the brand to cater to a broader range of driver preferences and skill levels.

As the Gallardo's production run progressed, the coexistence of AWD and RWD models showcased Lamborghini's commitment to offering choices to its discerning clientele. It demonstrated an understanding that while some customers prioritized ultimate performance and all-weather capability, others placed a premium on a more traditional, purist-oriented driving experience. This strategy not only broadened the Gallardo's appeal but also set the stage for future models to offer similar drivetrain options. This practice continues in Lamborghini's lineup to this day.

The story of the Gallardo's transmission and drivetrain evolution is more than just a tale of technological advancement. It's a narrative that reflects the changing demands and preferences of supercar buyers in the 21st century, as well as Lamborghini's ability to balance innovation with tradition. Whether equipped with all-wheel drive or rear-wheel drive, each Gallardo offered a unique driving experience, united by the unmistakable character of Lamborghini's acclaimed V10 engine.

Section 5.4: Launch Control and Performance Enhancements

The Lamborghini Gallardo's evolution wasn't just about raw power; it was also about harnessing that power effectively. One of the most significant advancements in this regard was the introduction of launch control, a feature that revolutionized the way drivers could accelerate from a standing start.

Launch control made its debut in later E-Gear-equipped Gallardos, marking a significant leap forward in performance technology. This system allowed drivers to achieve blistering acceleration from a standstill, consistently and reliably. The technology behind launch control was a complex dance of electronic systems working in harmony with the car's mechanical components.

At its core, the launch control system precisely managed clutch engagement and gear shifts to maximize acceleration while minimizing drivetrain stress. When activated, the system would hold the engine at an optimal RPM, typically around 5,000 revolutions per minute. Upon release of the brake pedal, the clutch would engage with carefully calculated slip, allowing the tires to maintain traction while transferring maximum power to the ground. Simultaneously, the transmission would execute rapid, perfectly timed shifts to keep the engine in its power band throughout the acceleration run.

As the Gallardo evolved, so did its launch control system. By the time the final Gallardo models rolled off the production line, the system had been refined to propel the car from 0 to 60 mph in just 3.2 seconds, a feat that would have seemed almost impossible when the first Gallardo was introduced in 2003.

But launch control wasn't the only performance-enhancing transmission feature introduced to the Gallardo over its lifespan. Later models featured a 'Thrust Mode' that optimized gear shifts for maximum acceleration at high speeds. This mode adjusted shift points and clutch engagement to minimize power loss during gear changes, allowing the Gallardo to accelerate with relentless force even at speeds where aerodynamic drag becomes a significant factor.

Another notable advancement was the introduction of a 'Corsa' mode in later E-Gear-equipped models. This track-focused setting sharpened throttle response, stiffened the magnetic ride suspension, and programmed the transmission for ultra-quick shifts. In Corsa mode, gear changes were executed with brutal efficiency, with shift times reduced to mere fractions of a second.

These transmission innovations had a profound impact on the Gallardo's performance reputation. While the car had always been respected for its power and handling, these advanced features helped cement its status as a technologically advanced, high-performance supercar. The Gallardo was no longer just about raw speed; it was about intelligent, efficient use of power.

Moreover, these advancements played a crucial role in making the Gallardo more accessible to a broader range of drivers. The sophisticated launch control and performance modes allowed less experienced drivers to extract professional-level performance from the car, while still providing the engagement and excitement that enthusiasts craved.

The evolution of these performance-enhancing features in the Gallardo also foreshadowed the direction of the broader supercar industry. Today, launch control, intelligent transmission modes, and other electronic aids are standard features on most high-performance vehicles, a trend that the Gallardo helped pioneer.

In essence, the Gallardo's transmission innovations represented a perfect synthesis of Lamborghini's traditional focus on performance with cutting-edge automotive technology. These features not only enhanced the car's capabilities but also transformed the driving experience, allowing drivers to push the limits of performance with greater confidence and control. As we reflect on the Gallardo's legacy, it's clear that these transmission advancements played a crucial role in shaping the modern supercar landscape.

Section 5.5: Transmission Reliability and Maintenance

The Lamborghini Gallardo's transmission systems, while innovative and performance-oriented, were not without their challenges when it came to reliability and maintenance. This section examines the intricacies of owning and maintaining a Gallardo, with a particular focus on its transmission systems.

When comparing the E-Gear and manual transmissions, reliability emerges as a key differentiator. The E-Gear system, while offering lightning-fast shifts and cutting-edge technology, generally required more frequent maintenance and was more prone to issues than its manual counterpart.

The traditional six-speed manual transmission, with its mechanical simplicity, proved to be incredibly robust and reliable over time. Many Gallardo enthusiasts argue that the manual transmission's durability significantly outweighs the performance benefits of the E-Gear system, especially for those who plan on owning their Gallardo for the long term.

However, E-Gear-equipped Gallardos weren't without their merits. When properly maintained, these systems could provide years of trouble-free operation. The key was adhering to a strict maintenance schedule and addressing any issues promptly. For instance, E-Gear systems require fluid changes every 15,000 miles, a crucial service that, if neglected, could lead to more serious problems down the line.

E-Gear-equipped Gallardos often relied heavily on their hydraulic systems, which were a common source of issues, particularly in earlier models. Failures in the hydraulic pump were not uncommon and could result in an inability to shift gears, rendering the vehicle immobile. Another frequent issue involved worn clutch packs, as the rapid gear changes facilitated by the E-Gear system could accelerate wear on these components. Although repairs for these problems could be expensive, they were generally predictable and could often be avoided with proper maintenance and care.

Manual transmission Gallardos, while generally more reliable, weren't entirely immune to issues. Clutch wear was the most common concern, particularly in cars that saw frequent track use or aggressive driving. However, clutch replacements in manual Gallardos were typically less expensive and less complex than addressing E-Gear-related issues.

For those looking to enhance their Gallardo's performance, the aftermarket offered a variety of upgrades for both transmission types. E-Gear-equipped cars could benefit from software updates that improved shift times and smoothness. For manual cars, upgraded clutches and lightweight flywheels were popular modifications, particularly among owners seeking to enhance engine power.

When considering a used Gallardo, the transmission type and its maintenance history should be at the forefront of any potential buyer's mind. A well-maintained E-Gear system can still provide an exhilarating driving experience, but a neglected one can quickly become a financial burden. On the other hand, a manual Gallardo with a documented service history can be a relatively worry-free ownership experience, provided the clutch is in good condition.

Ultimately, both transmission options in the Gallardo require respect and care. Regular maintenance, regardless of transmission type, is crucial. This includes not only fluid changes but also inspections of components such as shift forks, synchros, and, in the case of E-Gear cars, the hydraulic system. Owners who stay proactive with maintenance often report more positive long-term experiences, regardless of their transmission choice.

It's worth noting that as Gallardos age, finding skilled technicians familiar with their unique transmission systems, particularly the E-Gear, can become challenging. Building a relationship with a reputable Lamborghini specialist can be invaluable for long-term ownership.

In conclusion, while the Gallardo's transmission options each came with their own set of maintenance considerations, they both contributed to the car's legendary status. Whether one prefers the cutting-edge technology of the E-Gear or the raw, mechanical feel of the manual, proper care and maintenance are key to enjoying the Gallardo's performance for years to come. The choice between E-Gear and manual often comes down to personal preference, driving style, and willingness to manage potential maintenance costs, but both can provide an unforgettable driving experience when correctly cared for.

Section 5.6: The Legacy of Gallardo's Transmission Innovations

The Lamborghini Gallardo's journey through transmission technologies left an indelible mark on the supercar industry, influencing not only future Lamborghini models but also setting benchmarks for competitors. This legacy extends far beyond the Gallardo's production run, shaping the expectations of supercar enthusiasts and pushing the boundaries of what's possible in high-performance vehicles.

One of the most significant impacts of the Gallardo's transmission evolution was the normalization of paddle-shift systems in supercars.

Lamborghini Gallardo: A Decade of Domination

When the E-Gear was first introduced, it represented a bold step into the future of automotive technology. By the end of the Gallardo's production, paddle-shift transmissions had become the industry standard, with even the most traditional manufacturers adopting similar systems. The Gallardo played a crucial role in this shift, demonstrating that automated manual transmissions could deliver both blistering performance and an engaging driving experience.

The Gallardo's transmission innovations also had a profound effect on Lamborghini's subsequent models. The lessons learned and technologies developed for the Gallardo paved the way for even more advanced systems in cars like the Huracán and Aventador. For instance, the Huracán's dual-clutch transmission, while a significant leap forward, built upon the foundation laid by the Gallardo's E-Gear system. This technological lineage exemplifies how the Gallardo served as a crucial stepping stone in Lamborghini's ongoing pursuit of performance excellence.

Moreover, the Gallardo's offering of both automated and manual transmissions throughout much of its lifespan highlighted a critical debate in the automotive world: the balance between technological advancement and driver engagement. By providing both options, Lamborghini acknowledged the diverse preferences of its customer base, from those seeking cutting-edge technology to purists who valued the traditional manual experience. This approach influenced other manufacturers to maintain manual options in their high-performance vehicles for longer than they might have otherwise, preserving choice for enthusiasts.

The transmission developments in the Gallardo also played a significant role in democratizing supercar performance. The E-Gear system, with its launch control and lightning-fast shifts, made it possible for a broader range of drivers to experience the full potential of a Lamborghini.

This accessibility factor contributed to Gallardo's commercial success and helped broaden the appeal of supercars beyond the realm of experienced drivers. In the realm of motorsports, the Gallardo's transmission technologies found their way onto the racetrack.

The learnings from the road car's development influenced the creation of race-specific variants, enhancing Lamborghini's competitiveness in GT racing. This cross-pollination between road and race technology became a hallmark of Lamborghini's approach, further cementing the brand's performance credentials.

The Gallardo's transmission journey also sparked innovation among aftermarket tuners and performance shops. As enthusiasts sought to extract even more performance from their Gallardos, a whole industry emerged around upgrading and modifying these transmissions. This aftermarket support not only extended the performance envelope of the Gallardo but also contributed to its long-term desirability and value retention.

Lastly, the Gallardo's transmission evolution serves as a case study in the rapid pace of automotive technology advancement. From its introduction to its final production year, the Gallardo witnessed and participated in a period of unprecedented change in how power is delivered to the wheels of a supercar. This rapid evolution mirrors the broader technological leaps occurring in the automotive industry, from electrification to autonomous driving systems.

In conclusion, the legacy of the Gallardo's transmission innovations extends far beyond the car itself. It represents a pivotal chapter in the history of supercars, influencing industry trends, shaping driver expectations, and pushing the boundaries of performance. As we look to the future of high-performance vehicles, the groundwork laid by the Gallardo continues to inform and inspire, ensuring its place not just in Lamborghini's history but in the annals of automotive engineering.

Section 5.7: The Legacy of Gallardo's Transmission Innovations

The Lamborghini Gallardo's journey through transmission technologies left an indelible mark on the supercar landscape, influencing not only future Lamborghini models but the entire high-performance automotive industry. As we reflect on the Gallardo's decade-long production run, it's clear that its transmission innovations played a crucial role in shaping the car's identity and success.

The introduction of the E-Gear system in the early Gallardo models marked Lamborghini's bold step into the world of automated manual transmissions. This technology, which seemed almost revolutionary at the time, offered drivers the thrill of lightning-fast gear changes without sacrificing the raw, mechanical feel that Lamborghini enthusiasts craved. As the E-Gear system evolved, becoming smoother and more responsive with each iteration, it set a new standard for what drivers could expect from a modern supercar.

Simultaneously, Lamborghini's decision to continue offering a traditional manual transmission option throughout most of the Gallardo's lifespan demonstrated a deep understanding of its customer base. This move catered to purists who valued the intimate connection between car and driver that only a manual gearbox could provide. The Gallardo thus became a bridge between the old and new worlds of supercar engineering, appealing to a wide range of driving enthusiasts.

The Gallardo's transmission developments also played a significant role in the car's performance evolution. The implementation of launch control and other performance-enhancing features showcased how advanced transmission technology could dramatically improve real-world performance. These innovations helped the Gallardo remain competitive in an increasingly crowded supercar market, consistently delivering exhilarating performance that belied its long production run.

Perhaps most importantly, the Gallardo's transmission journey reflected and influenced broader trends in the automotive industry. The gradual shift in customer preference from manual to automated transmissions, clearly evident in Gallardo sales figures over the years, foreshadowed the near-extinction of manual transmissions in high-performance cars. Yet, the passionate following retained by manual Gallardos also contributed to a renewed appreciation for traditional transmissions, influencing decisions to offer manual options in some modern supercars.

The legacy of the Gallardo's transmission innovations can be seen in subsequent Lamborghini models and beyond. The lessons learned from the E-Gear system directly informed the development of the Huracán's dual-clutch transmission. At the same time, the expertise gained in all-wheel-drive integration continues to benefit Lamborghini's current lineup. Moreover, the Gallardo's influence extends to other manufacturers, who have sought to emulate the balance of performance, technology, and driver engagement that defined the Gallardo's transmission options.

In conclusion, the Gallardo's transmission story is one of evolution, innovation, and a deep understanding of the supercar market. It showcases Lamborghini's ability to push technological boundaries while respecting tradition. This balancing act contributed significantly to the Gallardo's status as one of the most successful and influential supercars of its era. As we move into an age of increasing electrification and automation in the automotive world, the lessons learned from the Gallardo's transmission journey continue to

Lamborghini Gallardo: A Decade of Domination

resonate, reminding us of the importance of driver engagement and the thrill of control in the supercar experience.

Chapter 6: Form Meets Function: Aerodynamics and Styling Changes

Section 6.1: The Initial Design - Form and Function in Harmony

When the Lamborghini Gallardo first burst onto the scene in 2003, it represented a masterful blend of form and function. The design team at Lamborghini, led by Luc Donckerwolke, had a clear mission: to create a supercar that was not only visually stunning but also aerodynamically efficient. This delicate balance between aesthetics and performance would become a hallmark of the Gallardo throughout its production run.

The original Gallardo's design philosophy was rooted in Lamborghini's tradition of bold, aggressive styling, but with a modern twist. The car's wedge-shaped profile and low-slung body were not merely for show; they played a crucial role in reducing drag and improving high-speed stability. This marriage of form and function was evident in every curve and angle of the Gallardo's bodywork.

Key aerodynamic features were seamlessly integrated into the Gallardo's striking design. The car's flat underbody and rear diffuser worked in tandem to create a ground effect, enhancing stability at high speeds. This feature, borrowed from racing technology, allowed the Gallardo to remain planted on the road without resorting to large, unsightly wings or spoilers that might disrupt its clean lines.

The front of the car featured carefully designed air intakes that not only fed cool air to the powerful V10 engine but also channeled airflow around the body to reduce drag. These intakes, along with the sharp, angular headlights, gave the Gallardo its distinctive and aggressive face, instantly recognizable as a Lamborghini.

One of the most iconic styling cues of the Gallardo was its Y-shaped LED daytime running lights. This design element, which would later be adopted across the Lamborghini range, served both aesthetic and functional purposes. The unique light signature made the Gallardo immediately identifiable, even from a distance, while also improving visibility and safety.

The development of the Gallardo's aerodynamics was not left to chance. Extensive wind tunnel testing at Lamborghini's facility in Sant'Agata Bolognese played a crucial role in refining the car's body shape. Engineers spent countless hours fine-tuning every surface, resulting in subtle refinements that optimized the Gallardo's drag coefficient without compromising its striking appearance.

The initial public and critical reception of the Gallardo's design was overwhelmingly positive. Automotive journalists praised the car for its perfect blend of aggression and elegance. Car and Driver magazine, in their first review of the Gallardo, called it "a benchmark for supercar aesthetics," highlighting how Lamborghini had managed to create a design that was both beautiful and functional.

Enthusiasts and potential buyers were equally impressed. The Gallardo's design struck a chord with those who wanted a supercar that could turn heads on the street but also deliver exceptional performance on the track. Its more compact dimensions compared to its larger sibling, the Murciélago, made it more accessible and usable as a daily driver, further broadening its appeal.

The success of the initial Gallardo design laid a strong foundation for the model's future development. It set a high bar for the balance between aesthetics and aerodynamics, a standard that Lamborghini would strive to surpass with each subsequent iteration of the Gallardo.

As we'll see in the following sections, this initial design would evolve over the years, incorporating new technologies and responding to changing regulations and market demands. However, the core principle of harmonizing form and function, so brilliantly executed in the original Gallardo, would remain a guiding light throughout the model's decade-long production run.

Section 6.2: The First Major Facelift - Refining the Formula

In 2008, five years after the Gallardo's initial release, Lamborghini unveiled the first major facelift for its popular supercar. This update was a calculated move to sharpen the car's looks and improve its aerodynamic efficiency, ensuring the Gallardo remained competitive in the rapidly evolving supercar market.

The most noticeable changes came to the front fascia of the vehicle. The revised front bumper featured larger air intakes, serving a dual purpose: to enhance the Gallardo's aggressive appearance while also improving cooling for the potent V10 engine. These enlarged intakes weren't just for show; they allowed for better airflow to critical components, helping to maintain optimal performance even under demanding conditions.

At the rear, the modifications were equally significant. The redesigned rear diffuser was a highlight of the facelift, increasing downforce by an impressive 31%. This substantial improvement in aerodynamics significantly enhanced the car's high-speed stability, allowing drivers to push the Gallardo to its limits with greater confidence. The revised rear also included subtle changes to the taillights and exhaust outlets, further refining the car's aesthetics while serving functional purposes.

The facelift wasn't limited to the front and rear; it also introduced new wheel designs that contributed to both the car's visual appeal and its aerodynamic performance. The new 'Apollo' wheel design, for instance, not only refreshed the Gallardo's side profile but also improved brake cooling through better airflow. This attention to detail demonstrated Lamborghini's commitment to enhancing every aspect of the car's design.

The impact of these styling changes on performance metrics was noteworthy. Wind tunnel tests showed that the facelifted Gallardo achieved a 3% reduction in drag coefficient. At the same time, this might seem like a small number; in the world of high-performance automobiles, such an improvement is significant. This reduction in drag translated to a top speed increase of 4 mph, a testament to the power of refined aerodynamics.

Beyond the performance gains, the facelift also served to modernize the Gallardo's appearance. The sharper lines and more aggressive stance brought the design more in line with contemporary supercar aesthetics, ensuring the Gallardo remained a head-turner on the streets and a desirable object for enthusiasts and collectors alike.

The interior also received attention during this facelift, with upgraded materials and a revised instrument cluster. These changes, while not directly related to aerodynamics, complemented the exterior modifications to create a more cohesive and up-to-date package.

Importantly, this facelift wasn't just about cosmetic changes. It represented Lamborghini's philosophy of continuous improvement, incorporating lessons learned from the first five years of Gallardo production. The changes were a result of extensive customer feedback, racing experience, and technological advancements, all aimed at enhancing the overall Gallardo experience.

The facelifted Gallardo was well-received by both the automotive press and customers. Critics praised the more aggressive look and the tangible performance improvements, while sales figures demonstrated that the updates had successfully reinvigorated interest in the model.

This major facelift set the stage for the latter half of the Gallardo's production run, introducing design elements and aerodynamic features that would be further refined in subsequent special editions and the final Gallardo models. It proved that even a successful design could be improved upon, and showcased Lamborghini's commitment to keeping its products at the cutting edge of supercar design and performance.

Section 6.3: Aerodynamic Innovations in Special Editions

The Lamborghini Gallardo's production run was punctuated by a series of special editions that not only showcased the brand's creativity but also served as testbeds for aerodynamic innovations. These limited-run models pushed the boundaries of what was possible with the Gallardo platform, often introducing cutting-edge aerodynamic features that would later influence the regular production models.

The Gallardo Superleggera, introduced in 2007, was a prime example of how special editions could dramatically enhance aerodynamic performance. This lightweight variant featured a large rear wing and a revised front splitter that together increased downforce by an impressive 120% compared to the standard model. The Superleggera's aggressive aerodynamic package not only improved high-speed stability but also contributed to its blistering lap times on racetracks around the world.

Not all special editions focused solely on performance, however. The Valentino Balboni Edition, named after Lamborghini's legendary test driver, introduced a unique styling element that subtly enhanced aerodynamics. The distinctive center stripe that ran the length of the car wasn't just a visual flourish; it was carefully designed to optimize airflow over the car's roof, demonstrating how even seemingly decorative elements could serve an aerodynamic purpose.

The racing world provided valuable lessons for road car aerodynamics, as evidenced by the Gallardo Super Trofeo. This track-only version of the Gallardo featured an extreme aerodynamic package that was then adapted for road use. The aggressive rear diffuser design from the Super Trofeo race car found its way onto the road-going Gallardo Superleggera, significantly enhancing high-speed stability and cornering grip. The most significant aerodynamic leap came with the introduction of the Gallardo Performante. This model marked Lamborghini's first foray into active aerodynamics on the Gallardo platform.

The Performante introduced electronically controlled flaps that could adjust airflow in real-time, optimizing the balance between drag reduction and downforce generation based on driving conditions. This technology represented a quantum leap in the Gallardo's aerodynamic capabilities, allowing for unprecedented levels of performance and efficiency.

The innovations introduced in these special editions didn't remain isolated to limited production runs. The success of the Superleggera's aerodynamic package, for instance, led to the incorporation of a more aggressive front splitter on all Gallardo models from 2010 onwards. Similarly, the active aerodynamic concepts explored in the Performante would go on to influence future Lamborghini models, setting the stage for even more advanced systems in cars like the Huracán and Aventador.

These special editions served a dual purpose: they satisfied the desire of Lamborghini enthusiasts for exclusive, high-performance variants, while also allowing the company to experiment with advanced aerodynamic concepts in a real-world setting. The lessons learned from these limited-run models were invaluable, driving the continuous improvement of the Gallardo's aerodynamic performance throughout its production life.

Moreover, the aerodynamic advancements made in these special editions helped Lamborghini maintain the Gallardo's competitiveness in an increasingly crowded supercar market. As rivals introduced new models with cutting-edge aerodynamics, Lamborghini was able to respond quickly by incorporating technologies and designs proven in its special edition models.

In essence, the special edition Gallardos were more than just collector's items or marketing exercises. They were rolling laboratories that pushed the boundaries of aerodynamic design, striking a balance between the need for downforce and stability, and the desire for speed and efficiency. The innovations born from these models not only enhanced the Gallardo's performance but also laid the groundwork for future Lamborghini supercars, ensuring that the brand remained at the forefront of automotive aerodynamics.

Section 6.4: The Final Facelift - Perfecting the Formula

In 2012, Lamborghini unveiled the final facelift for the Gallardo, marking the culmination of a decade's worth of refinement and innovation. This last update aimed to modernize the design while paying homage to the car's enduring legacy. The timing was crucial, as the Gallardo approached the end of its production run, and Lamborghini sought to give it a fitting send-off while bridging the gap to its eventual successor.

The front-end design underwent significant changes, introducing a more angular bumper reminiscent of the Aventador. This update not only improved aerodynamics but also strengthened the family resemblance across Lamborghini's model range. The revised front fascia featured larger air intakes, enhancing both the car's aggressive aesthetic and its cooling efficiency. These changes weren't merely cosmetic; they represented a careful balance between form and function, a hallmark of Lamborghini's design philosophy.

At the rear, the modifications had an equally significant impact. A redesigned diffuser and a subtle lip spoiler were introduced, resulting in an 8% increase in downforce compared to the previous model. This enhancement significantly improved high-speed stability, allowing drivers to push the Gallardo to its limits with greater confidence. The rear lights were also updated, incorporating LED technology for improved visibility and a more modern appearance.

One of the most notable changes in this final facelift was the full integration of LED lighting technology. The Gallardo became one of the first supercars to feature full LED headlights as standard. This update not only modernized the car's face but also improved nighttime visibility and reduced energy consumption. The distinctive Y-shaped LED daytime running lights, which had become a Lamborghini signature, were refined further, enhancing the car's unmistakable presence on the road.

The overall aerodynamic package of the Gallardo saw significant refinement in this final iteration. Wind tunnel testing revealed a 5% improvement in overall aerodynamic efficiency compared to the original 2003 model. This was achieved through a series of subtle but effective tweaks to the body shape, including a more efficiently sculpted hood, refined side skirts, and the aforementioned rear improvements. These changes not only enhanced the car's performance but also its fuel efficiency, a growing concern even in the supercar segment.

Lamborghini also took this opportunity to introduce new wheel designs, which played a dual role in enhancing both aesthetics and performance. The new wheels were not only lighter but also designed to improve brake cooling, a critical factor in maintaining consistent performance during high-speed driving or track use. Inside the car, the updates were more subtle but no less significant. The interior saw refinements in material quality and technology integration, with a new infotainment system that brought the Gallardo up to date with the latest in-car technologies. This balance of performance enhancement and luxury refinement was crucial in maintaining the Gallardo's position as a daily-drivable supercar.

The final facelift also introduced new color options and customization possibilities, allowing buyers to personalize their Gallardo to an even greater degree. This move acknowledged the growing trend of personalization in the luxury car market and ensured that even in its twilight years, each Gallardo could be as unique as its owner.

Perhaps most importantly, this final update to the Gallardo demonstrated Lamborghini's commitment to continuous improvement. Even as they were undoubtedly deep in development of the Gallardo's successor, the company invested significant resources in refining a model nearing the end of its life cycle. This dedication to excellence, regardless of a product's stage in its lifecycle, is a testament to Lamborghini's philosophy and a key reason for the Gallardo's enduring appeal.

The 2012 facelift, therefore, wasn't just a cosmetic update but a comprehensive refinement of the Gallardo formula. It represented the culmination of a decade's worth of learning, innovation, and response to customer feedback. This final iteration of the Gallardo stood as a worthy capstone to a model that had redefined Lamborghini as a brand and set new standards in the supercar segment. As the curtain began to fall on the Gallardo's remarkable run, this last update ensured it would leave the stage at the peak of its powers, its legacy secure as one of the greatest and most influential supercars of its era.

Section 6.5: Materials and Manufacturing: Advancing Aerodynamics

The evolution of materials and manufacturing techniques played a crucial role in advancing the Lamborghini Gallardo's aerodynamic performance throughout its production run. As technology progressed, Lamborghini engineers and designers were able to push the boundaries of what was possible in terms of both form and function.

One of the most significant advancements came with the increased use of carbon fiber in body panels. By 2013, the Gallardo Superleggera utilized carbon fiber for 70% of its body panels, marking a substantial increase from earlier models. This shift not only reduced the overall weight of the vehicle but also allowed for more complex aerodynamic shapes to be incorporated into the design. The lightweight nature of carbon fiber allowed for the addition of larger aerodynamic elements, such as wings and diffusers, without significantly impacting the car's weight distribution or center of gravity.

Advancements in aluminum chassis technology also played a pivotal role in the Gallardo's aerodynamic evolution. Lamborghini's improvements in aluminum extrusion techniques resulted in a 20% increase in chassis rigidity without adding weight. This enhanced structural integrity allowed for more precise aerodynamic tuning, as the stiffer chassis could better withstand the forces generated by high-speed airflow. The improved rigidity also meant that aerodynamic elements could be more finely tuned, as there was less flex in the body structure to account for.

As the Gallardo's production run progressed, Lamborghini began integrating active aerodynamic elements into the design. The Gallardo Performante, for instance, introduced electronically controlled air intakes that could open or close based on the car's cooling needs. This real-time optimization of airflow not only improved engine performance but also allowed for better management of the car's overall aerodynamic profile depending on driving conditions.

Improvements in manufacturing precision also contributed significantly to the Gallardo's aerodynamic advancements. By 2013, Lamborghini's use of computer-controlled milling for mold creation allowed for body panel tolerances of less than 0.1mm. This level of precision was crucial for optimal aerodynamic performance, as even small gaps or misalignments in body panels can create turbulence and increase drag at high speeds. The tight tolerances ensured that the carefully designed aerodynamic surfaces functioned as intended, maximizing the car's performance potential.

One of the most influential factors in the Gallardo's aerodynamic development was the role of computational fluid dynamics (CFD) in the design process. Advanced CFD simulations allowed Lamborghini engineers to test hundreds of subtle design variations virtually, optimizing the Gallardo's aerodynamics to a degree impossible with physical prototypes alone. This technology enabled the team to fine-tune every curve and surface of the car, predicting and minimizing areas of turbulence or high pressure that could negatively impact performance.

The use of CFD also allowed for more rapid iteration in the design process. Engineers could quickly test and refine ideas without the need for costly and time-consuming physical prototypes. This accelerated development cycle meant that aerodynamic improvements could be implemented more frequently and effectively throughout the Gallardo's lifespan.

Moreover, the insights gained from CFD simulations often led to counterintuitive design choices that proved highly effective in real-world conditions. For example, subtle changes to the angle of the A-pillars or the curvature of the side mirrors, which might seem insignificant to the naked eye, could result in meaningful reductions in drag or improvements in downforce when tested virtually and subsequently validated in wind tunnel tests.

The combination of advanced materials, precision manufacturing, and sophisticated design tools allowed Lamborghini to continuously refine the Gallardo's aerodynamic package throughout its production run. From the sleek lines of the original 2003 model to the highly optimized final editions, each iteration of the Gallardo benefited from these technological advancements.

This constant evolution not only improved the car's performance metrics but also contributed to its enduring aesthetic appeal. The

Gallardo's ability to remain fresh and competitive in a rapidly advancing market was due in no small part to Lamborghini's commitment to leveraging cutting-edge materials and manufacturing techniques in pursuit of aerodynamic excellence.

Section 6.6: The Impact of Regulations on Design

As the Lamborghini Gallardo evolved over its decade-long production run, it faced an ever-changing landscape of automotive regulations. These regulations, designed to improve safety, reduce emissions, and mitigate noise pollution, presented significant challenges to Lamborghini's designers and engineers. The task at hand was to maintain the Gallardo's iconic styling and exhilarating performance while adhering to increasingly stringent standards.

One of the most visible impacts of regulations on the Gallardo's design came from pedestrian safety standards, particularly in Europe. The 2008 facelift introduced a slightly raised hood line to comply with new European pedestrian safety regulations. This change posed a significant challenge for Lamborghini's designers, who had to find creative ways to maintain the car's low-slung, aggressive appearance while meeting the new requirements. The solution involved subtle reshaping of the front end, resulting in a design that satisfied both regulatory demands and aesthetic expectations.

Emissions regulations also played a crucial role in shaping the Gallardo's aerodynamics over time. As governments worldwide tightened restrictions on vehicle emissions, Lamborghini had to find ways to improve the car's fuel efficiency without compromising its performance. This led to increased focus on aerodynamic efficiency, particularly in the later models. The final Gallardo iterations featured improved underbody aerodynamics, which reduced drag and improved fuel efficiency by 3%. These changes, although not immediately visible to the casual observer, were crucial in helping the

Gallardo meet stringent emissions standards while maintaining its status as a supercar.

Noise regulations presented another unique challenge for the Gallardos' design team. The distinctive, high-pitched wail of the V10 engine was a key part of the car's character, but it also had to comply with increasingly strict noise limits. To address this, engineers redesigned the rear diffuser to act as an additional sound dampener. This innovative solution allowed the Gallardo to maintain its signature sound while meeting regulatory requirements, all without compromising the car's aerodynamic performance.

The world of motorsport also influenced the Gallardo's road car design, particularly as racing regulations evolved. When new GT3 regulations were introduced in 2013, Lamborghini had to adapt the Gallardo Super Trofeo race car to comply. The aerodynamic lessons learned from this process led to the development of a more efficient rear wing for the final road car models. This transfer of technology from track to street not only improved the Gallardo's performance but also strengthened its racing pedigree.

Throughout its lifespan, one of the most significant challenges faced by Lamborghini was balancing regulatory compliance with the Gallardo's design identity. Despite the numerous regulatory hurdles, Lamborghini's designers managed to preserve the Gallardo's distinctive wedge profile and aggressive stance. They achieved this by creatively integrating required safety features into the car's aerodynamic package. For instance, the redesigned front bumper for pedestrian safety was sculpted to enhance airflow, turning a potential compromise into an aerodynamic advantage.

The impact of regulations on the Gallardo's design wasn't limited to external features. Interior design also had to evolve to meet safety standards, with changes to airbag placement and the use of materials. These interior modifications were carried out with the same attention to detail and design flair that characterized the exterior modifications, ensuring that the Gallardo's cockpit remained a perfect blend of luxury and sportiness.

As the automotive industry moved towards electrification, even the last iterations of the Gallardo were influenced by this trend. While the Gallardo never became a hybrid, considerations for future powertrain options influenced some of the late-stage design decisions, laying the groundwork for preplanning the hybrid technology that would be adopted as the successor, the Huracán.

In conclusion, the evolution of the Lamborghini Gallardo's design in response to regulations is a testament to the ingenuity and adaptability of Lamborghini's engineers and designers. They successfully navigated a complex regulatory landscape while maintaining the essence of what made the Gallardo a true Lamborghini. This delicate balance between compliance and design integrity not only kept the Gallardo relevant throughout its production run but also set important precedents for future Lamborghini models in an increasingly regulated automotive world.

Section 6.7: The Gallardo's Design Legacy

The Lamborghini Gallardo's impact on automotive design extends far beyond its decade-long production run. As a pivotal model for the Italian automaker, the Gallardo's evolution in styling and aerodynamics left an indelible mark on both Lamborghini's future lineup and the broader supercar industry.

The Gallardo's influence is immediately apparent in subsequent Lamborghini models. Its signature Y-shaped LED lighting and sharp, angular body lines became defining characteristics of the brand's design language. This aesthetic DNA is clearly visible in the

Gallardo's successor, the Huracán, as well as the flagship Aventador. The Gallardo effectively set the template for modern Lamborghini design, balancing aggressive styling with aerodynamic efficiency in a way that has become synonymous with the brand.

Beyond Lamborghini, the Gallardo's design philosophy sparked a trend across the supercar world. Competitors like Ferrari and McLaren took note of the Gallardo's successful marriage of form and function, incorporating similar approaches in their own designs. The Gallardo demonstrated that a car could be both visually striking and aerodynamically efficient, a concept that has since become a cornerstone of supercar design.

Several aerodynamic technologies pioneered or refined on the Gallardo have become industry standards. The use of a flat underbody and rear diffuser for improved downforce, while not entirely new, was perfected on the Gallardo and subsequently adopted widely across the high-performance car segment. These features, once considered exotic, are now expected elements in any serious performance vehicle.

The Gallardo's role in shaping public perception of supercar aesthetics cannot be overstated. Its distinctive silhouette became so iconic that it defined a generation's idea of what a supercar should look like. This influence extended beyond the automotive world, impacting popular culture through its appearances in movies, video games, and even toy car designs. The Gallardo's look became shorthand for "supercar" in the public imagination.

Perhaps most importantly, the lessons learned from the Gallardo's decade-long evolution informed Lamborghini's approach to future models. The experience gained in aerodynamic development, material use, and design refinement directly influenced the creation of the Huracán. This resulted in a 50% improvement in aerodynamic efficiency for the Huracán compared to the original Gallardo, demonstrating the value of the knowledge accumulated over the Gallardo's lifespan.

The Gallardo also taught Lamborghini valuable lessons about longevity in design. The car's ability to remain visually compelling and performance-relevant over a ten-year production run, with only two major facelifts, showed that a well-conceived initial design could have staying power. This insight has influenced Lamborghini's approach to model lifecycles and update strategies.

Furthermore, the Gallardo's success in balancing performance, style, and everyday usability set a new benchmark for Lamborghini. It proved that a Lamborghini could be both an extreme performance machine and a relatively practical daily driver. This philosophy has carried forward into all of the brand's subsequent models.

In conclusion, the Gallardo's design legacy is multi-faceted and far-reaching. It shaped Lamborghini's future, influenced industry trends, set new standards for aerodynamic performance, and redefined public expectations of supercar design. As we look at modern supercars, the echoes of the Gallardo's groundbreaking design are still clearly visible, a testament to its enduring impact on the automotive world.

Lamborghini Gallardo: A Decade of Domination

Chapter 7: Special Editions: Limited Run Gallardo Models

Section 7.1: The Art of Exclusivity: Lamborghini's Approach to Special Editions

Lamborghini has long understood the allure of exclusivity in the automotive world. The creation of special edition models is not merely a marketing tactic but a philosophy deeply ingrained in the brand's DNA. For Lamborghini, limited editions serve as a canvas for pushing the boundaries of design, performance, and luxury, while simultaneously igniting the passions of car enthusiasts and collectors worldwide.

At the core of Lamborghini's philosophy behind creating limited editions is the desire to offer something truly unique to its most discerning customers. These special models often feature bespoke design elements, enhanced performance capabilities, and exclusive materials that set them apart from their standard counterparts. By doing so, Lamborghini not only caters to the individual tastes of its clientele but also reinforces its position as a leader in automotive innovation and craftsmanship.

The role of special editions in brand prestige and marketing cannot be overstated. These exclusive models serve as halo products, drawing attention to the entire Lamborghini lineup and elevating the brand's overall image. They generate buzz in automotive circles, dominate headlines, and create a sense of anticipation among enthusiasts. This heightened visibility and desirability translate into increased brand value, helping to maintain Lamborghini's status as a premier luxury sports car manufacturer.

Throughout the Gallardo's decade-long production run, limited editions played a crucial role in the model's evolution. Each special edition served as a testbed for new technologies, design concepts, and performance enhancements. Many of the innovations first introduced in these exclusive variants eventually found their way into the standard Gallardo lineup, contributing to the continuous improvement and refinement of the model over its lifespan.

The impact of limited editions on resale value and collectibility is significant. Due to their rarity and unique features, special edition Gallardos often command a premium in the used car market. Collectors view these models as prime investment opportunities, with many limited-run variants appreciating in value over time. This phenomenon not only benefits the original owners but also reinforces Lamborghini's reputation as a maker of highly desirable and valuable automobiles.

Lamborghini's approach to special editions often involves collaboration with renowned designers and other prestigious brands. These partnerships bring fresh perspectives and expertise to the table, resulting in truly distinctive creations. Whether it's working with fashion houses to develop bespoke interiors or teaming up with aerospace companies to incorporate cutting-edge materials, these collaborations push the boundaries of what's possible in automotive design and engineering.

The art of exclusivity as practiced by Lamborghini through its special editions is a delicate balance of innovation, prestige, and market savvy. By consistently delivering exceptional and limited-run models, Lamborghini not only satisfies the desires of its most passionate customers but also continually reinforces its position at the forefront of the supercar world. The Gallardo, as we will explore in the following sections, became a perfect canvas for this approach, with numerous special editions that captivated the automotive world and left an indelible mark on the brand's legacy.

Section 7.2: Gallardo SE (Special Edition) - The Pioneer

The Lamborghini Gallardo SE, introduced in 2005, marked the beginning of a new era for the Italian automaker's approach to limited-run models. As the first special edition of the Gallardo lineup, the SE set the stage for future exclusive variants and played a crucial role in elevating the model's status among supercar enthusiasts.

The Gallardo SE was a masterclass in subtle yet impactful design modifications. Lamborghini chose to distinguish this model with a two-tone exterior color scheme, pairing the body color with a striking black roof. This visual contrast immediately sets the SE apart from its standard counterparts, giving it a more aggressive and sporty appearance. The exterior was further enhanced by a set of specially designed "Callisto" wheels, which not only looked stunning but also improved the car's overall aerodynamics.

Inside the cabin, the Gallardo SE continued to impress with its attention to detail. The interior featured a unique two-tone color scheme that mirrored the exterior, creating a cohesive design language throughout the vehicle. Premium leather upholstery adorned the seats, dashboard, and door panels, while carbon fiber accents added a touch of motorsport-inspired aesthetics. A special plaque denoting the car's limited-edition status and production number was prominently displayed, reminding occupants of the vehicle's exclusivity.

Performance-wise, the Gallardo SE received several enhancements over the standard model. The 5.0-liter V10 engine was tuned to produce an additional 10 horsepower, bringing the total output to an impressive 520 hp. This power boost, combined with a recalibrated six-speed e-gear transmission, resulted in improved acceleration and a more responsive driving experience. The SE also featured a sports-tuned suspension setup, providing sharper handling and better cornering capabilities without sacrificing the Gallardo's renowned everyday usability.

Lamborghini limited the production of the Gallardo SE to just 250 units worldwide, instantly making it a highly sought-after collector's item. The market reception was overwhelmingly positive, with all units selling out rapidly despite a significant price premium over the standard Gallardo. This success demonstrated the appetite for exclusive, limited-run supercars and validated Lamborghini's strategy of creating special editions.

The long-term impact of the Gallardo SE on future special editions cannot be overstated. It established a template for Lamborghini to follow, showing how subtle design changes, performance enhancements, and limited availability could create a highly desirable product. The SE's success paved the way for more daring and experimental special editions in the future, allowing Lamborghini to push the boundaries of design and engineering while maintaining the Gallardo's core appeal.

Moreover, the Gallardo SE helped solidify the model's position in the competitive supercar market. By offering a more exclusive version of an already popular car, Lamborghini was able to attract a new subset of buyers who sought the prestige of owning a limited-edition vehicle. This strategy not only boosted sales but also enhanced the overall brand image, positioning Lamborghini as a manufacturer that could cater to the most discerning automotive enthusiasts.

In retrospect, the Gallardo SE represents a pivotal moment in the Gallardo's history. It marked the beginning of a series of special editions that would each leave their unique mark on the automotive world. The SE's blend of subtle design modifications, performance enhancements, and exclusivity set a high bar for future limited-run models, not just for Lamborghini but for the entire supercar industry. As the pioneer of Gallardo special editions, the SE holds a special place in the hearts of Lamborghini aficionados and continues to be a highly prized collector's item to this day.

Section 7.3: Gallardo Nera - The Dark Knight

The Lamborghini Gallardo Nera, unveiled at the 2006 Paris Motor Show, stands as a testament to the allure of darkness in automotive design. This limited edition model, aptly named "Nera" (Italian for "black"), was a bold statement that pushed the boundaries of luxury and exclusivity within the Gallardo lineup.

The concept behind the Nera edition was rooted in Lamborghini's desire to create a Gallardo that exuded mystery and sophistication. Drawing inspiration from the sleek, shadowy aesthetics often associated with stealth and power, the design team set out to craft a vehicle that would captivate onlookers with its commanding presence.

The most striking feature of the Gallardo Nera was, undoubtedly, its distinctive black-themed design elements. The exterior was adorned with a special matte black paint, a finish that was still relatively uncommon in the automotive world at the time. This unique paint job not only set the Nera apart from its glossy counterparts but also gave it a subtle, almost predatory appearance. The effect was further enhanced by the contrast provided by select body parts finished in bright chrome, including the side mirrors, rear pillars, and parts of the front bumper.

The Nera's dark theme wasn't limited to its exterior. Step inside, and you'd find yourself enveloped in a cocoon of luxury that continued the black motif. The interior was a masterpiece of contrasts, featuring black leather upholstery with white stitching that accentuated the cabin's contours. This was complemented by carbon fiber trim elements, adding a high-tech feel to the already futuristic interior.

Lamborghini didn't stop at aesthetics when creating the Nera. The company also focused on using exclusive materials to elevate the driving experience. The seats were crafted from a combination of leather and Alcantara, providing both comfort and grip during spirited driving. The steering wheel was wrapped in perforated leather, offering improved grip and tactile feedback to the driver.

In terms of performance, the Gallardo Nera retained the standard model's formidable 5.0-liter V10 engine, producing 520 horsepower. However, the driving experience was subtly enhanced by a specially tuned suspension system, which provided a slightly more aggressive ride without compromising the Gallardo's everyday usability. The result was a car that not only looked the part of a menacing supercar but also delivered an exhilarating driving experience to match its appearance.

From a collector's perspective, the Gallardo Nera holds a special place in Lamborghini's history. With only 185 units produced, it represents a rare and highly sought-after variant of the Gallardo. Its limited production run, combined with its unique aesthetic approach, has made it a prized possession among Lamborghini enthusiasts and collectors alike.

The Nera edition also played a crucial role in shaping Lamborghini's approach to special editions. Its success demonstrated the market's appetite for highly individualized, limited-run models, paving the way for future special editions across Lamborghini's lineup.

In essence, the Gallardo Nera was more than just a black-painted Gallardo; it was a distinct model. It was a bold experiment in automotive design, pushing the boundaries of what a luxury sports car could be. By embracing darkness not as an absence of color, but as a powerful aesthetic in its own right, Lamborghini created a vehicle that continues to captivate and inspire long after its initial release. The Gallardo Nera truly lived up to its unofficial moniker - "The Dark Knight" of the supercar world.

Section 7.4: Gallardo Superleggera - The Lightweight Champion

The Lamborghini Gallardo Superleggera represents the pinnacle of lightweight performance engineering within the Gallardo lineup. Introduced in 2007, the Superleggera, which translates to "super light" in Italian, was Lamborghini's answer to the growing demand for a more track-focused version of their popular supercar.

The development of the Superleggera concept was rooted in Lamborghini's racing heritage and the automotive industry's constant pursuit of better power-to-weight ratios. Engineers at Sant'Agata Bolognese set out to create a Gallardo that would not only outperform its standard counterpart but also challenge the best from rival manufacturers.

Weight reduction was at the heart of the Superleggera's philosophy. Lamborghini employed a range of advanced materials and innovative techniques to shed kilograms wherever possible. Carbon fiber, a material synonymous with high-performance vehicles, was extensively used throughout the car. The engine cover, rear diffuser, side mirrors, and various interior components were crafted from this lightweight yet strong material. Even the wheel nuts were made of titanium to save precious grams.

The result of this fanatical approach to weight saving was impressive. The Superleggera tipped the scales at 100 kg (220 lbs) less than the standard Gallardo, a significant reduction that had a profound impact on the car's performance and handling characteristics.

But Lamborghini didn't stop at weight reduction. The 5.0-liter V10 engine received a series of modifications to extract more power. Revised engine management software, a more free-flowing exhaust system, and other tweaks increased the output to 530 horsepower, a modest but meaningful bump over the standard Gallardo's 520 hp.

The combination of reduced weight and increased power transformed the Gallardo's performance. The Superleggera could sprint from 0 to 100 km/h (62 mph) in just 3.8 seconds, shaving precious tenths off the standard car's time. More importantly, the car's power-to-weight ratio improved significantly, enhancing acceleration across the entire speed range.

Aerodynamics played a crucial role in the Superleggera's performance envelope. The front and rear bumpers were redesigned to improve airflow and increase downforce. A larger rear diffuser and a fixed rear wing worked in tandem to keep the car planted at high speeds. These aerodynamic enhancements not only improved straight-line stability but also increased cornering speeds, allowing drivers to carry more momentum through turns.

On the track, the Superleggera truly came into its own. The reduced weight made the car more agile and responsive, allowing for quicker direction changes and later braking points. The enhanced aerodynamics provided drivers with increased confidence at high speeds, while the extra power ensured blistering acceleration out of corners.

When compared to its competitors, the Gallardo Superleggera held its own against the likes of the Ferrari F430 Scuderia and the Porsche 911 GT3 RS. Its lap times at various circuits around the world were consistently impressive, cementing its status as an actual track weapon.

The Superleggera wasn't just about numbers and lap times, though. It offered a more visceral and engaging driving experience than the standard Gallardo. The lighter weight amplified every input, making the car feel more alive and connected to the driver. The enhanced engine note, courtesy of the revised exhaust system, added to the sensory overload, creating an unforgettable experience for anyone lucky enough to get behind the wheel.

The Gallardo Superleggera represented a significant milestone in the model's evolution. It showcased Lamborghini's ability to push the boundaries of performance while staying true to the Gallardo's core attributes. This lightweight champion not only elevated the Gallardo's status among enthusiasts but also paved the way for future high-performance variants across Lamborghini's lineup.

The lessons learned from the development of the Superleggera would influence future models, including the Aventador and Huracán. Its legacy lives on in Lamborghini's continued commitment to lightweight construction and track-focused performance, ensuring that the spirit of the Superleggera continues to inspire the brand's engineering philosophy.

Section 7.5: Gallardo LP 550-2 Valentino Balboni - A Tribute to a Legend

The Lamborghini Gallardo LP 550-2 Valentino Balboni edition stands as a testament to the profound impact one individual can have on an automotive brand. This special edition not only honors a legendary figure but also marks a significant departure from Lamborghini's established norms, offering enthusiasts a unique driving experience that harks back to the company's roots.

Valentino Balboni's story is intrinsically linked with Lamborghini's history. Hired personally by Ferruccio Lamborghini in 1967, Balboni started as an apprentice mechanic and eventually became the company's chief test driver. Over his illustrious 40-year career, Balboni played a pivotal role in developing and refining nearly every Lamborghini model, earning him an almost mythical status among enthusiasts and within the company itself.

What sets the LP 550-2 Valentino Balboni edition apart is its unique rear-wheel-drive configuration. This marked a significant departure from Lamborghini's standard all-wheel-drive setup, which had been a hallmark of the brand since the introduction of the Diablo VT in 1993. The decision to create a rear-wheel-drive Gallardo was not just a technical choice, but a philosophical one, aimed at delivering a more pure, engaging driving experience - one that Balboni himself preferred.

The rear-wheel-drive layout fundamentally altered the car's dynamics. With 550 horsepower directed solely to the rear wheels, the LP 550-2 offered a more challenging and rewarding driving experience. It demanded more skill from the driver, harking back to the era of Lamborghini's earlier models that Balboni had helped develop. This configuration resulted in a car that was slightly lighter than its all-wheel-drive counterparts, further enhancing its agility and responsiveness.

Visually, the Balboni edition featured several bespoke design elements that distinguished it from others. A white and gold stripe ran the length of the car, a nod to the colors of Balboni's helmet.

The interior featured a unique "Polar White" leather with green, white, and red accents - the colors of the Italian flag - paying homage to the car's Italian heritage. Each vehicle also bore a plaque with Balboni's signature, cementing its status as a true collector's item. The driving dynamics of the LP 550-2 Valentino Balboni were met with enthusiasm by purists and critics alike.

The rear-wheel-drive setup allowed for a level of driver engagement that many felt had been missing from modern Lamborghinis. It offered a rawer, more visceral experience that was closer to the spirit of classic Lamborghinis. The car's ability to perform controlled drifts and power slides - a feat much more challenging in all-wheel-drive models - was particularly praised.

The legacy of the Balboni edition extends far beyond its limited production run of 250 units. It represented a pivotal moment in Lamborghini's modern history, showcasing the company's willingness to look back at its roots while pushing forward with innovation. The success and positive reception of this model paved the way for future rear-wheel-drive Lamborghinis, including subsequent Gallardo models and even influencing the development of the Huracán RWD.

Moreover, the Balboni edition served as a fitting tribute to a man who had dedicated his life to Lamborghini. It encapsulated Valentino

Balboni's philosophy of driving enjoyment and his invaluable contributions to the brand. In many ways, this special edition Gallardo became a rolling museum piece, preserving a crucial part of Lamborghini's heritage and the legacy of one of its most beloved figures.

The Gallardo LP 550-2 Valentino Balboni edition remains highly sought after by collectors and enthusiasts. Its unique configuration, limited production numbers, and the story behind its creation have cemented its place as one of the most special and significant Gallardo models ever produced. It stands as a shining example of how a limited edition can be more than just a marketing exercise, but a genuine celebration of automotive passion and heritage.

Section 7.6: Gallardo LP 570-4 Squadra Corse - Track-Focused Excellence

The Lamborghini Gallardo LP 570-4 Squadra Corse represented the pinnacle of track-focused engineering within the Gallardo lineup. Drawing inspiration from Lamborghini's successful racing program, this limited edition model bridged the gap between road-going supercars and their track-only counterparts.

At the heart of the Squadra Corse's development was a desire to bring the thrill of motorsport to the streets. Lamborghini's racing division, Squadra Corse, lent not only its name but also its expertise to create a Gallardo that could dominate both road and track. The result was a car that pushed the boundaries of what was possible in a street-legal vehicle.

The most striking aspect of the Squadra Corse was its array of track-oriented modifications. The car featured a revised aerodynamic package, including a large rear wing and redesigned front and rear diffusers. These changes weren't merely cosmetic; they significantly increased downforce, providing enhanced stability at high speeds and through corners. The wing, in particular, was derived directly from the Gallardo Super Trofeo race car, highlighting the model's racing pedigree.

Lamborghini's engineers also focused on weight reduction, a crucial factor in improving track performance. The extensive use of carbon fiber for body panels and interior components resulted in a weight saving of 70 kg compared to the standard Gallardo LP 560-4. This weight reduction, combined with the uprated 570-horsepower V10 engine, resulted in a power-to-weight ratio that rivaled that of many purpose-built race cars.

The interior of the Squadra Corse reflected its track-focused nature. Gone were many of the luxury amenities found in other Gallardo models, replaced by a stripped-down, driver-centric cockpit. Racing bucket seats, crafted from carbon fiber and clad in Alcantara, held the driver firmly in place during high-G maneuvers.

The steering wheel was redesigned for optimal grip and feedback, while the instrument cluster was simplified to provide only the most critical information at a glance. Performance-wise, the Squadra Corse was a force to be reckoned with. It could accelerate from 0 to 100 km/h in just 3.4 seconds, with a top speed of 320 km/h.

However, straight-line speed was just part of the equation. The real magic of the Squadra Corse was in its handling. The combination of reduced weight, improved aerodynamics, and a retuned suspension resulted in a car that was incredibly agile and responsive, capable of carving through corners with precision and speed that left even experienced drivers in awe.

Comparing the Squadra Corse to the standard Gallardo revealed just how far Lamborghini had pushed the envelope. Lap times at various tracks showed improvements of several seconds, a significant margin in the world of high-performance cars. The Squadra Corse wasn't just faster; it was more engaging, more communicative, and more rewarding to drive at the limit.

Despite its track-focused nature, the Squadra Corse remained road-legal, allowing owners to drive their race-bred machines to and from the track. This duality made it particularly appealing to enthusiasts who wanted the ultimate expression of the Gallardo's performance potential without the logistical challenges of owning a dedicated race car.

The Gallardo LP 570-4 Squadra Corse represented more than just the ultimate track-focused Gallardo; it was a statement of Lamborghini's engineering prowess and a testament to the Gallardo platform's incredible versatility. It showcased how a production car could be transformed into something truly extraordinary, blurring the lines between road car and race car in a way that few others could match. For many, the Squadra Corse remains the ultimate expression of the Gallardo's potential, a fitting swansong for a model that had redefined the supercar landscape over its decade-long production run.

Section 7.7: Other Notable Special Editions

While the Gallardo's most famous special editions often stole the spotlight, several other limited-run models deserve recognition for their unique attributes and contributions to the Gallardo legacy. These editions not only showcased Lamborghini's creativity but also demonstrated the brand's ability to cater to specific markets and celebrate various themes.

The Gallardo Bicolore stands out with its striking two-tone paint scheme. This special edition, introduced in 2011, featured a bold contrast between the roof and pillars, painted in Noctis Black, and the rest of the body, available in a choice of five colors. This distinctive look gave the Gallardo a fresh aesthetic appeal, demonstrating that even subtle design changes could have a significant impact on the car's overall appearance.

For those seeking the ultimate in luxury and exclusivity, Lamborghini created the Gallardo LP 560-4 Gold Edition. Limited to just ten units for the Chinese market, this model epitomized opulence with its unique gold paint and matching gold-trimmed interior. The Gold Edition demonstrated Lamborghini's willingness to push the boundaries of luxury, creating a truly one-of-a-kind supercar that appealed to collectors and enthusiasts alike.

Lamborghini also recognized the importance of catering to specific markets with models like the Gallardo LP 550-2 Hong Kong Edition. Created to celebrate the 20th anniversary of Lamborghini Hong Kong, this rear-wheel-drive special edition featured a unique Bianco Monocerus white exterior with red and gold stripes, colors inspired by the Bauhinia Blakeana flower on Hong Kong's flag. Only eight units were produced, making it one of the rarest Gallardo models ever created.

In a similar vein, the Gallardo Malaysia Limited Edition paid homage to the Malaysian market. This special model, based on the LP 550-2 model, sported a striking Matte Blue Caelum exterior color, complemented by a cabin upholstered in black Alcantara with blue stitching. Limited to just 20 units, this edition highlighted Lamborghini's commitment to its global markets and ability to create bespoke models that resonated with local audiences.

Lastly, the Gallardo Tricolore deserves mention for its celebration of Italian heritage. Created to commemorate the 150th anniversary of Italian unification, this special edition featured a white exterior adorned with red, white, and green stripes running along the car's length, a nod to the Italian flag.

The interior continued the patriotic theme with green, white, and red stitching on the seats, steering wheel, and dashboard. Limited to just 150 units, the Tricolore served as a rolling tribute to Italy's rich history and Lamborghini's proud Italian roots.

These special editions, while perhaps not as widely known as some of their counterparts, played a crucial role in the Gallardo's story. They demonstrated Lamborghini's versatility in creating unique models that appealed to different tastes, markets, and themes. From celebrating national pride to pushing the boundaries of luxury, these limited-run Gallardos showcased the model's adaptability and Lamborghini's commitment to exclusivity.

Moreover, these editions helped maintain interest in the Gallardo throughout its long production run, offering collectors and enthusiasts new and exciting variants to pursue. They also allowed Lamborghini to experiment with different design elements, color combinations, and themes, some of which would influence future models across the brand's lineup.

In the grand tapestry of Gallardo special editions, these models may have been produced in smaller numbers. Still, their impact on the Gallardo's legacy and their importance to Lamborghini's history should not be underestimated.

They represent the brand's ability to surprise, delight, and cater to the diverse tastes of its global clientele, further cementing the Gallardo's status as one of the most successful and versatile supercars of its generation.

Chapter 8: On the Track: The Gallardo's Racing Pedigree

Section 8.1: The Gallardos' Transition to Motorsports

The Lamborghini Gallardo's journey from a road-going supercar to a formidable force on the racetrack is a testament to the brand's commitment to performance and innovation. As Lamborghini sought to expand its presence in the motorsports world, the decision to enter the Gallardo in competitive racing marked a significant milestone for the company.

The transition from street to track was not without its challenges. Engineers faced the daunting task of adapting a car designed primarily for road use to withstand the rigors of competitive racing. This process involved a comprehensive overhaul of various systems, including the suspension, aerodynamics, and powertrain, to meet the demanding requirements of motorsports.

Lamborghini Gallardo: A Decade of Domination

The development of the first racing-specific Gallardo models was a meticulous process that involved countless hours of testing and refinement. Lamborghini's engineers worked tirelessly to extract every ounce of performance from the already potent platform, pushing the boundaries of what was possible with the Gallardo's V10 engine and all-wheel-drive system.

Collaboration with experienced racing teams and professional drivers played a crucial role in fine-tuning the Gallardo for competitive use. Their feedback was invaluable in identifying areas for improvement and developing solutions that would give the Gallardo an edge on the track. This partnership between Lamborghini's engineers and motorsports professionals ensured that the racing Gallardo would be competitive from the outset.

The debut of the Gallardo in professional motorsports was met with great anticipation and excitement. As the sleek Italian supercar lined up on the grid for the first time, it represented not just a new competitor but a bold statement from Lamborghini about its intentions in the world of racing. The Gallardos' first forays into competitive events were closely watched by enthusiasts and rivals alike, all eager to see how this newcomer would perform against established racing pedigrees.

From its early appearances, the Gallardo demonstrated that it was more than capable of holding its own on the track. Its combination of powerful performance, agile handling, and striking aesthetics made it a fan favorite and a serious contender in various racing series. As the Gallardo continued to evolve and improve with each race, it became clear that Lamborghini had successfully transformed its road-going supercar into a true motorsports contender.

The Gallardo's transition to motorsports not only enhanced its reputation but also provided valuable insights that would influence future road car development. The lessons learned on the track would go on to inform improvements in performance, handling, and durability across Lamborghini's entire range, creating a virtuous cycle of innovation between the company's road and race cars.

Section 8.2: The Gallardo GT3

The introduction of the Gallardo GT3 in 2006 marked a significant milestone in Lamborghini's racing history. This purpose-built race car was designed to compete in the highly competitive GT3 class, showcasing the Gallardo's potential as a formidable track weapon.

The Gallardo GT3 was a testament to Lamborghini's engineering prowess, featuring extensive modifications from its road-going counterpart. The car's body was stripped down and rebuilt with lightweight materials, including carbon fiber panels and a full roll cage, resulting in a weight reduction of over 300 kilograms. Aerodynamic enhancements were also implemented, with a large rear wing, front splitter, and redesigned side skirts to improve downforce and stability at high speeds.

Under the hood, the Gallardo GT3 retained the road car's 5.2-liter V10 engine, but it was tuned to produce over 520 horsepower. The power was channeled through a sequential six-speed gearbox, specially designed for racing applications. The suspension system was completely overhauled, featuring adjustable dampers and race-spec springs to provide better handling and cornering ability on various track surfaces.

One of the most significant advantages of the Gallardo GT3 on the track was its excellent power-to-weight ratio. The combination of increased power output and substantial weight reduction resulted in blistering acceleration and improved agility through corners. The car's mid-engine layout also contributed to its exceptional balance and handling characteristics, making it a favorite among professional drivers.

The Gallardo GT3's success in various championships worldwide quickly established its reputation as a serious contender in the GT3 class. It claimed numerous victories and podium finishes in prestigious events, including the FIA GT3 European Championship, the Italian GT Championship, and the Asian Le Mans Series. The car's reliability and consistent performance made it a popular choice for both factory-backed teams and private racing outfits.

As the GT3 model evolved throughout its production run, Lamborghini continued to refine and improve its performance. Subsequent iterations saw further aerodynamic enhancements, engine upgrades, and chassis improvements. These developments not only kept the Gallardo GT3 competitive against newer rivals but also provided valuable insights that would influence future Lamborghini road and race cars.

The Gallardo GT3's success on the track had a profound impact on Lamborghini's brand image and racing program. It demonstrated that the Italian manufacturer could produce not just stunning road cars but also competitive race machines capable of going toe-to-toe with established motorsport brands. This success laid the foundation for Lamborghini's expanded involvement in GT racing and paved the way for future models to continue the legacy of the Gallardo GT3.

In essence, the Gallardo GT3 represented the perfect fusion of Lamborghini's road car expertise and racing ambitions. It proved that the Gallardo platform was not just a stylish supercar for the streets, but a versatile and potent base for a world-class racing machine. The GT3's achievements on tracks around the globe cemented the Gallardo's place in motorsports history and set a new standard for Lamborghini's future racing endeavors.

Section 8.3: Super Trofeo Series

The Lamborghini Blancpain Super Trofeo series, launched in 2009, marked a significant milestone in the Gallardo's racing legacy. This one-make championship was conceived as a platform for gentleman drivers and professional racers alike to experience the thrill of Lamborghini racing in a competitive yet controlled environment.

At the heart of the Super Trofeo series was the Gallardo LP 560-4 Super Trofeo, a race-prepared version of the road-going supercar. This track-focused machine boasted an array of enhancements that set it apart from its street-legal counterpart. The 5.2-liter V10 engine was tuned to produce 570 horsepower, a modest increase over the standard model, but the real transformation lay in its chassis and aerodynamics.

The Super Trofeo featured a stripped-out interior, a full roll cage, and a race-spec suspension system that drastically improved its on-track performance. Aerodynamic modifications included a large rear wing, a front splitter, and various other bodywork changes designed to increase downforce and improve cooling. These alterations resulted in a car that was not only faster around a circuit but also more predictable and more manageable to drive at the limit.

The Super Trofeo championship was structured to provide an engaging and accessible racing format. Typically, each event consisted of free practice sessions, qualifying, and two races. The series initially launched in Europe but quickly expanded to include Asian and North American championships, reflecting the global appeal of both the Gallardo and Lamborghini's racing program.

Throughout its run, the Super Trofeo series saw numerous memorable races and produced several standout champions. The close racing and relatively level playing field ensured that skill and strategy often prevailed over outright car performance, leading to thrilling on-track battles and nail-biting championship conclusions.

One of the most notable aspects of the Super Trofeo series was its impact on Lamborghini's brand and customer racing program. The championship served as a powerful marketing tool, showcasing the performance capabilities of Lamborghini vehicles in a high-octane, competitive environment. It allowed wealthy enthusiasts to experience the thrill of racing in a factory-backed series, fostering a sense of community and brand loyalty among Lamborghini's most devoted customers.

Moreover, the Super Trofeo acted as a proving ground for both car development and driver talent. Many of the lessons learned from the series directly influenced the evolution of the Gallardo road car, particularly in terms of handling, aerodynamics, and driver interface. For drivers, the championship provided a stepping stone to higher levels of GT racing, with several Super Trofeo alums going on to achieve success in international GT3 competition.

The Gallardo Super Trofeo series also played a crucial role in establishing Lamborghini's long-term commitment to motorsport. Its success paved the way for the expansion of Lamborghini's racing activities, including the development of more advanced GT3 cars and the continuation of the Super Trofeo concept with the Gallardo's successor, the Huracán.

In essence, the Super Trofeo series transformed the Gallardo from a formidable road car into a true racing icon. It provided a platform for Lamborghini to showcase its engineering prowess, engage with its most passionate customers, and firmly establish its presence in the world of modern motorsport.

The legacy of the Gallardo Super Trofeo continues to influence Lamborghini's racing programs to this day, cementing its place as a pivotal chapter in the marque's illustrious racing history.

Section 8.4: Endurance Racing Achievements

The Lamborghini Gallardo's prowess on the racetrack extended far beyond short sprint races, as it proved its mettle in the grueling world of endurance racing. These long-distance events, often spanning 24 hours or more, pushed the limits of both man and machine, testing reliability, efficiency, and stamina alongside raw speed.

The Gallardo's foray into endurance racing began with its participation in some of the most prestigious events on the motorsport calendar. Among these, the 24 Hours of Spa and the 24 Hours of Nürburgring stood out as true tests of the car's capabilities. At Spa-Francorchamps, the legendary Belgian circuit known for its challenging layout and unpredictable weather, the Gallardo showcased its ability to maintain consistent performance over extended periods. Teams fielding Gallardo GT3 models often found themselves battling for class victories, with several podium finishes to their credit.

The Nürburgring 24 Hours, held on the infamous Nordschleife, presented an even greater challenge. Known as the "Green Hell," this track's combination of long straights, tight corners, and significant elevation changes made it the ultimate proving ground for any race car. The Gallardo's participation in this event was a testament to Lamborghini's confidence in its durability and performance.

To adapt the Gallardo for these punishing races, significant modifications were necessary. Engineers focused on enhancing reliability without sacrificing the car's inherent speed. This included optimizing the cooling systems to handle the increased stress of prolonged high-speed running, developing more durable brake components to withstand hours of heavy use, and fine-tuning the suspension to maintain consistent handling characteristics as fuel loads changed throughout the race.

Perhaps the most crucial adaptation was in the realm of aerodynamics. While the Gallardo's sleek design was already optimized for performance, endurance racing demanded even greater efficiency. Advanced aerodynamic packages were developed, incorporating larger rear wings, more aggressive front splitters, and refined underbody designs. These modifications not only improved downforce for better cornering speeds but also enhanced overall efficiency, crucial for managing fuel consumption over long race distances.

The challenges faced in endurance events were numerous and varied. Tire management became a critical factor, with teams and drivers having to balance outright pace with the need to extend tire life. The Gallardo's weight distribution and suspension setup played a crucial role in minimizing tire wear, enabling more consistent performance over extended periods.

Another significant challenge was managing the car's systems over such extended periods. The Gallardo's advanced electronics and powertrain components were pushed to their limits, requiring meticulous preparation and real-time management during the races. Teams developed sophisticated monitoring systems to track every aspect of the car's performance, enabling preemptive maintenance and informed decision-making during pit stops.

Lamborghini Gallardo: A Decade of Domination

Weather often played a crucial role in these endurance events, particularly at circuits like Spa and the Nürburgring. The Gallardo's all-wheel-drive system proved to be a significant advantage in changing conditions, providing drivers with confidence and consistency regardless of whether the track was dry, damp, or entirely wet.

Throughout its endurance racing career, the Gallardo faced stiff competition from established manufacturers with decades of experience in long-distance racing. Yet, it consistently proved its worth, often punching above its weight against more specialized prototypes and purpose-built endurance racers.

The legacy of the Gallardo in endurance racing extends beyond mere results. It demonstrated Lamborghini's commitment to building not just fast cars, but robust, reliable machines capable of performing at the highest level for extended periods. This experience directly influenced the development of future Lamborghini models, both on the track and for the road.

Moreover, the Gallardo's endurance racing exploits helped to dispel any lingering notions that Lamborghinis were purely style-over-substance machines. It proved that beneath the stunning Italian design lay a car of genuine substance, capable of taking on the most demanding challenges in motorsport.

In the annals of endurance racing history, the Lamborghini Gallardo carved out its own chapter, one marked by impressive performances, technological advancements, and a spirit of relentless pursuit of excellence. Its achievements in these grueling events not only enhanced the model's reputation but also played a crucial role in shaping Lamborghini's future direction in both racing and road car development.

Section 8.5: Track-Focused Road Models

The Lamborghini Gallardo's racing success didn't just stay on the track; it found its way onto public roads through a series of track-focused models that brought the thrill of motorsports to everyday driving. These special editions not only showcased Lamborghini's racing prowess but also pushed the boundaries of what was possible in a road-legal supercar.

The journey began with the introduction of the Gallardo Superleggera in 2007. This lightweight version of the Gallardo was a direct response to Ferrari's Challenge Stradale, aiming to offer a more visceral driving experience. The Superleggera, Italian for "super light," lived up to its name by shedding 100 kg (220 lbs) through the extensive use of carbon fiber and other lightweight materials. This weight reduction, combined with a power increase to 530 hp, resulted in a car that was not just faster but also more agile and responsive on both road and track.

Building on the success of the Superleggera, Lamborghini took things a step further with the LP 570-4 Squadra Corse in 2013. This model represented the ultimate fusion of racing technology and road car practicality. Named after Lamborghini's racing department, the Squadra Corse was a road-going version of the Super Trofeo race car. It featured a 570-horsepower V10 engine, a rear wing that provided three times the downforce of the standard Gallardo, and a quick-release engine cover straight from the world of motorsports.

The performance enhancements in these track-focused models went beyond just power increases. Lamborghini's engineers employed sophisticated weight reduction techniques, such as using carbon fiber for the engine cover, rear wing, and large sections of the interior. The suspension was retuned for sharper handling, and carbon-ceramic brakes were fitted as standard, providing fade-free stopping power even under intense track use.

When compared to their standard counterparts, these track-focused models offered a significantly different driving experience. They were rawer, more responsive, and more communicative, providing a level of driver engagement that was closer to a race car than a typical road-going supercar. The Squadra Corse, for instance, could accelerate from 0-100 km/h in just 3.4 seconds and reach a top speed of 320 km/h (199 mph), figures that were remarkable even by supercar standards of the time.

The impact of these track-focused Gallardo models on the supercar market was profound. They set a new benchmark for performance and driver engagement, forcing competitors to respond with their own track-oriented variants. More importantly, they helped to establish Lamborghini as a brand that could credibly straddle the worlds of road cars and racing machines.

These models also played a crucial role in Lamborghini's marketing strategy. They served as halo products, showcasing the brand's technical capabilities and reinforcing its racing pedigree. The limited production numbers of these special editions also helped to increase their desirability and collectible status among enthusiasts.

The success of the track-focused Gallardo models paved the way for future Lamborghini special editions, influencing the development of models like the Huracán Performante and the Aventador SVJ. They demonstrated that there was a market for extreme, track-oriented supercars that could also be driven on the road. This philosophy continues to shape Lamborghini's product strategy to this day.

In essence, these track-focused road models represented the culmination of everything Lamborghini had learned from its racing endeavors with the Gallardo. They brought the excitement and technology of the racetrack to the road, allowing enthusiasts to experience a taste of Lamborghini's motorsport success in a road-legal package. These models not only enhanced the Gallardo's legacy but also set a new standard for what a road-going supercar could achieve.

Section 8.6: Technical Innovations Born from Racing

The Gallardo's racing career wasn't just about trophies and lap times; it was a crucible for technological advancement. The intense demands of competitive motorsports pushed Lamborghini's engineers to innovate constantly, resulting in a cascade of improvements that benefited both the racing models and their road-going counterparts.

Aerodynamics played a crucial role in the Gallardo's on-track success. Wind tunnel testing and real-world racing experience led to the development of more efficient front splitters, rear diffusers, and adjustable rear wings. These elements not only increased downforce for better cornering stability but also reduced drag for higher top speeds. The knowledge gained from sculpting the air around the racing Gallardos influenced the design of later road models, resulting in cars that were not only more striking but also more aerodynamically efficient.

The heart of any racing car is its engine, and the Gallardo's V10 powerplant underwent significant evolution thanks to its time on the track. Engineers fine-tuned the engine management systems, optimized fuel injection, and improved cooling to extract every last horsepower while maintaining reliability under the extreme conditions of endurance racing. These enhancements trickled down to production models, resulting in more powerful and responsive engines that could withstand the rigors of both track days and daily driving.

The Gallardo's suspension system saw substantial refinements born from countless laps around the world's most demanding circuits. Racing teams experimented with various spring rates, damper settings, and anti-roll bar configurations to find the perfect balance between stability and responsiveness. This knowledge led to the development of more sophisticated adaptive suspension systems in later road-going Gallardos, offering drivers a wider range of performance characteristics at the touch of a button.

Braking systems underwent a revolution during the Gallardo's racing career. The extreme temperatures and forces experienced during competitive driving led to the development of more efficient cooling systems for the brake rotors and calipers. Carbon-ceramic brakes, once the exclusive domain of racing cars, made their way into high-performance road models, offering unparalleled stopping power and fade resistance.

One of the most significant areas of improvement was in the Gallardo's all-wheel-drive system. Racing provided invaluable data on how to optimize power distribution for maximum traction and handling. This led to more advanced electronic differentials and traction control systems that could adapt to various driving conditions in milliseconds, enhancing both performance and safety.

The racing program also accelerated the adoption of lightweight materials throughout the Gallardo's construction. Carbon fiber, already widely used in motorsports, has found its way into more components of road cars. From body panels to interior trim, these weight-saving measures improved performance across the board, from acceleration to fuel efficiency.

Data acquisition and telemetry systems, essential tools in racing, influenced the development of more advanced driver aids and information systems in production Gallardos. This not only improved the driving experience but also allowed for better monitoring of vehicle systems, enhancing reliability and maintenance. The intense focus on driver ergonomics in racing cockpits led to improvements in the road cars' interiors. Better-positioned controls, more supportive seats, and enhanced visibility all stemmed from lessons learned on the track.

Ultimately, the technical innovations born from the Gallardo's racing exploits created a virtuous cycle of improvement. Racing success drove customer interest, which in turn funded further development, leading to more advanced road cars and even more competitive racing machines. This symbiotic relationship between road and track solidified the Gallardo's position as not just a successful race car, but as a technological pioneer in the supercar world.

Section 8.7: The Gallardo's Racing Legacy

The Lamborghini Gallardo's racing legacy is a testament to its extraordinary performance and the dedication of the teams that campaigned it. Throughout its competitive career, the Gallardo amassed an impressive array of victories and championships across various racing categories, solidifying its place in motorsports history.

In GT3 racing, the Gallardo proved to be a formidable competitor, securing numerous wins in prestigious events such as the Spa 24 Hours and the Blancpain GT Series. Its success was not limited to Europe; the Gallardo also claimed victories in the Asian Le Mans Series and the Australian GT Championship, demonstrating its versatility and adaptability to different racing conditions and regulations.

Lamborghini Gallardo: A Decade of Domination

When compared to its competitors in the same racing categories, the Gallardo often stood out for its reliability and consistent performance. While other manufacturers struggled with reliability issues during endurance races, the Gallardo frequently crossed the finish line, earning a reputation for robustness that mirrored its road-going counterpart. This reliability, combined with its speed and agility, made it a favorite among privateer teams looking for a competitive yet dependable race car.

The Gallardo's racing program had a profound impact on Lamborghini's future motorsports endeavors. The success and lessons learned from the Gallardo's competitive career directly influenced the development of subsequent racing programs, most notably that of its successor, the Huracán. The experience gained in aerodynamics, engine performance, and chassis dynamics during the Gallardo's racing years provided invaluable data that Lamborghini's engineers could apply to future models.

Speaking of the Huracán, the influence of the Gallardo's racing pedigree is evident in its design and performance capabilities. The Huracán GT3, which replaced the Gallardo in Lamborghini's customer racing program, built upon the strengths of its predecessor while addressing areas for improvement identified during the Gallardo's racing career. This evolutionary approach ensured that Lamborghini remained at the forefront of GT racing, with the Huracán quickly establishing itself as a worthy successor on the track. In the broader context of Lamborghini's racing history, the Gallardo occupies a special place.

It marked the brand's return to factory-supported motorsports after a long hiatus and played a crucial role in reestablishing Lamborghini as a serious contender in modern sports car racing. The Gallardo's success on the track not only enhanced the brand's sporting credentials but also served as a powerful marketing tool, attracting a new generation of enthusiasts and customers to the Lamborghini brand.

Moreover, the Gallardo's racing program fostered a closer connection between Lamborghini and its customers. The Super Trofeo series, in particular, allowed gentleman drivers and racing enthusiasts to experience the thrill of competing in a Lamborghini, creating a community of passionate owners and racers that continues to support the brand's motorsports activities.

As we reflect on the Gallardos' legacy, it's clear that its impact extends far beyond the trophies and accolades it accumulated. The Gallardo's success on the track validated Lamborghini's engineering prowess, enhanced the brand's prestige, and laid the groundwork for future racing endeavors. It demonstrated that Lamborghini could produce not just stunning road cars but also competitive race machines capable of challenging and beating established motorsports brands.

In the annals of Lamborghini's history, the Gallardo will always be remembered as the car that brought the raging bull back to the racetrack, roaring with Italian passion and leaving an indelible mark on the world of motorsports. Its legacy continues to inspire and influence Lamborghini's approach to both road and race car development, ensuring that the spirit of the Gallardo lives on in every Lamborghini that takes to the track.

Chapter 9: Technology and Innovation: Electronics and Driver Aids

Section 9.1: The Foundation of Gallardo's Electronic Systems

The Lamborghini Gallardo's introduction in 2003 marked a significant leap forward in the integration of advanced electronics within the realm of high-performance supercars. As the first Lamborghini developed under Audi's ownership, the Gallardo benefited greatly from the German automaker's expertise in automotive electronics.

At the heart of the Gallardo's initial electronic architecture lay a sophisticated network of interconnected systems, designed to enhance performance, safety, and driver experience. This foundation was built upon a series of electronic control units (ECUs) that governed various aspects of the vehicle's operation. These ECUs were responsible for managing everything from engine performance and transmission control to chassis dynamics and driver aids.

Audi's influence on Lamborghini's electronic systems was profound. The Gallardo inherited Audi's advanced Controller Area Network (CAN bus) technology, which allowed for seamless communication between different electronic components. This integration enabled faster data transfer and more efficient management of the car's various systems, resulting in improved overall performance and responsiveness.

The first-generation Gallardo featured several key ECUs that formed the backbone of its electronic infrastructure. The Engine Control Module (ECM) was the most critical, overseeing the intricate operations of the V10 powerplant.

It worked in tandem with the Transmission Control Module (TCM) to optimize gear shifts and power delivery. The Anti-lock Braking System (ABS) and Electronic Stability Control (ESC) modules ensured that the Gallardo's immense power could be safely harnessed, even in challenging driving conditions.

Another crucial component of the Gallardo's electronic foundation was the implementation of drive-by-wire technology. This system replaced traditional mechanical linkages between the throttle pedal and the engine with electronic sensors and actuators. The result was more precise throttle control and the ability to integrate advanced features such as traction control and different driving modes.

When compared to its contemporary rivals, the Gallardo's electronic systems were at the forefront of supercar technology. While some traditionalists initially expressed concern about the potential for electronics to dilute the raw driving experience, the Gallardo proved that advanced technology could enhance, rather than detract from, the thrill of piloting a high-performance machine.

The sophistication of the Gallardo's electronic architecture allowed for continuous improvement throughout its production run. Software updates and hardware refinements could be implemented more easily, enabling Lamborghini to evolve the Gallardo's capabilities over time without the need for extensive mechanical redesigns.

This foundation of electronic systems laid the groundwork for future innovations in the Gallardo lineup. It provided the flexibility to introduce more advanced driver aids, improve performance parameters, and enhance the overall driving experience as technology progressed. The integration of these electronic systems not only set a new standard for Lamborghini but also influenced the direction of the entire supercar industry, demonstrating that cutting-edge technology and passionate engineering could coexist to create truly exceptional vehicles.

Section 9.2: Engine Management and Performance Electronics

The evolution of the Gallardo's engine management and performance electronics played a crucial role in its transformation from a raw, unbridled supercar to a sophisticated driving machine. At the heart of this evolution was the continuous development of the Gallardo's engine control module (ECM). With each iteration, the ECM became more advanced, capable of processing an increasing amount of data and making split-second adjustments to optimize engine performance.

One of the most significant advancements in the Gallardo's engine management was the implementation of drive-by-wire technology. This system replaced the traditional mechanical linkage between the accelerator pedal and the engine with an electronic connection. The result was a more precise and responsive throttle control, allowing drivers to modulate power delivery with greater accuracy. This technology not only improved drivability but also enabled the integration of more advanced driver aids and performance features.

The fuel injection and ignition control systems saw substantial improvements throughout the Gallardo's lifecycle. Early models utilized a relatively simple multi-point fuel injection system, but later versions incorporated more sophisticated direct injection technology. This advancement allowed for more precise fuel delivery, resulting in improved fuel efficiency and power output. Similarly, the ignition system evolved to provide more accurate spark timing, further enhancing performance and efficiency.

As the Gallardo matured, Lamborghini introduced variable valve timing electronics to the V10 engine. This system allowed for real-time adjustment of valve timing, optimizing performance across the entire rev range. At low speeds, it could be tuned for better fuel economy and smoother operation, while at high speeds, it maximized power output. The integration of this technology marked a significant leap forward in the Gallardo's ability to balance everyday drivability with supercar performance.

The culmination of these electronic advancements resulted in remarkable performance gains over the Gallardo's lifetime. The original 2003 model produced 493 horsepower, while the final 2013 Gallardo LP 570-4 Squadra Corse boasted an impressive 562 horsepower. This increase in power was not merely a result of mechanical improvements. Still, it was primarily due to the sophisticated engine management systems that could extract maximum performance while maintaining reliability and compliance with emissions regulations.

Moreover, these electronic systems allowed for different driving modes, a feature that became increasingly refined in later Gallardo models. Drivers could select between various preset configurations that altered throttle response, shift points, and other parameters to suit different driving conditions or preferences. This adaptability further showcased how electronics had transformed the Gallardo from a one-dimensional speed machine into a multi-faceted supercar capable of thrilling on the track and comfortable cruising on the street.

The advancements in engine management and performance electronics in the Gallardo not only improved its capabilities but also set new standards for the supercar industry. Competitors were forced to follow suit, leading to a new era where electronic sophistication became as important as raw horsepower in the supercar arms race. The Gallardo's journey from its relatively straightforward beginnings to its electronically enhanced final form mirrored the broader transformation of the automotive industry in the early 21st century, where the lines between mechanical engineering and computer science became increasingly blurred.

Section 9.3: Transmission and Drivetrain Electronics

The Lamborghini Gallardo's transmission and drivetrain systems underwent significant electronic enhancements throughout its production run, revolutionizing the way power was delivered to the wheels and how drivers interacted with the vehicle.

At the heart of this evolution was the development of the E-gear system's electronic controls. Initially introduced as an option in the first-generation Gallardo, the E-gear system represented a leap forward in automated manual transmission technology. The system utilized sophisticated electronics to control clutch engagement, gear selection, and shifting speed.

Lamborghini Gallardo: A Decade of Domination

Over time, Lamborghini's engineers refined the E-gear's electronic brain, resulting in faster shift times and smoother operation. By the late-model Gallardos, the E-gear system had become so advanced that it could execute gear changes in mere milliseconds, rivaling the speed of contemporary dual-clutch transmissions.

Equally important was the evolution of the all-wheel-drive system's electronic management. The Gallardo's Viscous Traction (VT) system, which could vary the torque split between the front and rear axles, relied heavily on electronic controls to optimize traction in various driving conditions. As the model progressed, these controls became more sophisticated, allowing for quicker and more precise adjustments to the power distribution. This resulted in improved handling characteristics and better overall performance, particularly in challenging weather conditions or during high-speed cornering.

One of the most exciting additions to the Gallardo's electronic arsenal was the implementation of launch control. This feature, which became available in later models, utilized a complex array of sensors and control units to optimize the vehicle's acceleration from a standing start. The system would hold the engine at the ideal RPM, manage clutch engagement, and precisely meter out power to all four wheels for the perfect launch. As with other systems, Lamborghini continually refined the launch control electronics, making it more effective and easier to use with each iteration.

The Gallardo also saw advancements in electronic differential controls and torque vectoring. These systems worked in concert with the all-wheel-drive electronics to distribute power not just between the front and rear axles, but also between the left and right wheels. This allowed for more precise control of the car's handling characteristics, enhancing both performance and safety. In later models, the torque vectoring system became so advanced that it could predictively adjust power distribution based on steering input and vehicle dynamics, further improving the Gallardo's already impressive cornering abilities.

When compared to its competitors, the Gallardo's transmission electronics often stood out for their sophistication and effectiveness. While some rivals relied on traditional manual transmissions or early dual-clutch systems, Lamborghini's E-gear and associated electronics offered a unique blend of performance and technology that appealed to both purists and tech enthusiasts alike.

The continuous evolution of the Gallardo's transmission and drivetrain electronics played a crucial role in maintaining its competitiveness throughout its long production run. These advancements not only improved the car's performance metrics but also enhanced its drivability and appeal to a broader range of customers. From the precision of the E-gear system to the intelligent power distribution of the all-wheel-drive, these electronic systems worked in harmony to deliver the thrilling yet controlled driving experience that became synonymous with the Gallardo name.

Section 9.4: Chassis and Suspension Electronics

The Lamborghini Gallardo's evolution in chassis and suspension electronics marked a significant leap forward in the supercar's handling capabilities and ride comfort. As the model progressed through its lifecycle, Lamborghini engineers continuously refined and introduced advanced electronic systems to enhance the vehicle's dynamic performance.

One of the most notable advancements came with the introduction of electronic damper control in later Gallardo models. This system allowed for real-time adjustment of the suspension's damping characteristics, providing an optimal balance between comfort and performance.

Drivers could select from multiple modes, ranging from a more compliant setting for everyday driving to a firmer, track-focused setup. The electronic damper control system utilized sensors throughout the vehicle to monitor road conditions, vehicle speed, and driver inputs, adjusting the damping force accordingly to maintain optimal tire contact with the road surface.

Building upon this technology, Lamborghini developed the magnetorheological suspension system for the Gallardo. This innovative system utilized magnetically controlled fluid in the dampers, enabling near-instantaneous changes in suspension stiffness. By applying an electromagnetic field to the fluid, its viscosity could be altered, providing a wider range of damping adjustments and faster response times compared to traditional electronic dampers. This technology not only improved handling performance but also enhanced ride comfort, adapting to road conditions more effectively than ever before.

The evolution of the Gallardo's Electronic Stability Control (ESC) system played a crucial role in enhancing the car's safety and performance. Early versions of the system focused primarily on preventing loss of control in emergencies. However, as the technology progressed, the ESC became more sophisticated, integrating with other vehicle systems to provide a more holistic approach to vehicle dynamics. Later iterations of the Gallardo's ESC worked in harmony with the all-wheel-drive system, traction control, and ABS to offer a more seamless and less intrusive driving experience, allowing skilled drivers to explore the car's limits while maintaining a safety net.

Traction control systems in the Gallardo also saw significant advancements throughout the model's lifespan. Early versions were relatively simple, cutting engine power or applying brakes to prevent wheel spin. As technology evolved, traction control became more nuanced, allowing for varying degrees of wheel slip depending on the selected driving mode. This progression culminated in a system that could optimize traction for different surfaces and driving conditions, from dry tarmac to wet roads, enhancing both performance and safety.

One of the most impressive achievements in the Gallardo's chassis electronics was the integration of cornering enhancements through electronic chassis management. This system used a combination of sensors and control units to monitor the vehicle's behavior during cornering. By precisely controlling the distribution of power between the wheels, adjusting the suspension, and even applying subtle brake force to individual wheels, the system could actively influence the car's cornering behavior. This resulted in improved turn-in response, reduced understeer, and overall more neutral handling characteristics, allowing drivers to carry more speed through corners with greater confidence.

The synergy between these various electronic systems transformed the Gallardo's driving dynamics over its lifetime. What began as a raw, uncompromising supercar evolved into a more sophisticated and accessible high-performance vehicle. The integration of advanced chassis and suspension electronics allowed the Gallardo to offer a broader range of capabilities, from comfortable grand touring to razor-sharp track performance, all without sacrificing the visceral excitement that Lamborghini was known for.

These advancements in chassis and suspension electronics not only improved the Gallardo's performance metrics but also enhanced its everyday usability and driver confidence. By providing a more adaptable and intelligent platform, Lamborghini succeeded in creating a supercar that could deliver extreme performance when demanded, while also offering a level of refinement and control that made it more accessible to a wider range of drivers. This technological evolution played a crucial role in maintaining the Gallardo's competitiveness throughout its production run and set the stage for future Lamborghini models to push the boundaries of automotive technology even further.

Section 9.5: Driver Aids and Safety Systems

The evolution of driver aids and safety systems in the Lamborghini Gallardo reflects the broader automotive industry's push towards enhanced safety and driver assistance. As the Gallardo matured over its decade-long production run, it incorporated increasingly sophisticated electronic systems designed to protect occupants and improve the driving experience.

At the heart of the Gallardo's safety systems was its advanced anti-lock braking system (ABS). The initial ABS technology in early Gallardo models was already state-of-the-art, but Lamborghini continued to refine it throughout the car's lifespan. Later iterations featured more advanced sensors and faster-acting hydraulic units, allowing for finer control of brake pressure. This resulted in shorter stopping distances and improved stability during emergency braking situations, particularly on wet or slippery surfaces.

The Gallardo's passive safety features also saw significant improvements over time. Early models were equipped with front and side airbags, but later versions incorporated more advanced restraint systems. These included knee airbags for the driver and passenger, as well as seat-mounted side airbags with improved deployment algorithms. The car's electronic control unit (ECU) constantly monitored various sensors to determine the severity of a collision and deploy the appropriate airbags with precision timing.

As the Gallardo evolved, Lamborghini introduced more convenience-oriented driver aids. One notable addition was the hill-hold assist feature, which prevented the car from rolling backward on inclines. This was particularly useful in manual transmission models, making hill starts easier and safer. The system would automatically engage the brakes for a few seconds after the driver released the brake pedal, giving ample time to apply the throttle and move forward without rolling back.

In later Gallardo models, Lamborghini introduced an adaptive cruise control system. This advanced feature used radar sensors to maintain a set distance from the vehicle ahead, automatically adjusting the Gallardo's speed to match traffic conditions. While some purists argued that such systems detracted from the raw driving experience expected of a Lamborghini, many owners appreciated the added convenience during long highway drives.

Parking a low-slung supercar like the Gallardo in tight spaces could be challenging, so Lamborghini gradually improved the car's parking assistance features. Early models relied solely on the driver's skill, but later versions introduced parking sensors that provided audible warnings as the vehicle approached obstacles.

The final iterations of the Gallardo even offered a rear-view camera system, displaying a clear view of the area behind the car on the infotainment screen. This not only made parking easier but also enhanced safety by reducing the risk of low-speed collisions.

Throughout its production run, the Gallardo's driver aids and safety systems were continuously refined to strike a balance between performance and protection. Lamborghini engineers worked tirelessly to ensure that these systems enhanced rather than hindered the driving experience. For instance, the stability control system was tuned to allow a degree of slip before intervening, preserving the car's dynamic character while still providing a safety net.

The implementation of these driver aids and safety systems in the Gallardo marked a significant shift in Lamborghini's approach to vehicle design. It demonstrated that even high-performance supercars could benefit from advanced electronic assistance without compromising their core appeal. This philosophy would continue to influence future Lamborghini models, setting a new standard for safety and technology in the supercar segment.

As the automotive industry moves towards increased automation and electrification, the groundwork laid by the Gallardo's safety systems and driver aids continues to influence Lamborghini's latest models. The lessons learned from balancing performance with safety in the Gallardo have proven invaluable in developing even more advanced systems for cars like the Huracán and Aventador, ensuring that Lamborghini remains at the forefront of automotive technology while maintaining its reputation for building some of the world's most exciting and driver-focused supercars.

Section 9.6: Infotainment and Connectivity

The Lamborghini Gallardo's journey through the realm of infotainment and connectivity mirrors the rapid advancement of consumer electronics in the 21st century. As the Gallardo evolved, so did its ability to keep drivers connected and entertained, transforming the supercar experience from purely visceral to technologically sophisticated.

In its early years, the Gallardo's audio system was relatively basic by today's standards. It featured a standard radio and CD player, with sound quality that was adequate but not exceptional. However, Lamborghini quickly recognized the importance of a premium audio experience in a high-end vehicle. As the model progressed, they introduced higher-quality speakers and amplifiers, culminating in an optional premium sound system in later iterations that rivaled those found in luxury sedans.

Navigation systems were not initially a priority for the Gallardo, as the focus was primarily on driving dynamics. However, as satellite navigation became more commonplace in luxury vehicles, Lamborghini integrated this technology into the Gallardo. Early systems were somewhat clunky and slow, but they improved significantly over time. Later models featured more responsive touchscreens, improved map accuracy, and faster route calculations, enhancing the Gallardo's practicality for both long trips and city driving.

The integration of Bluetooth technology marked a significant leap forward in the Gallardo's connectivity. Initially introduced for hands-free calling, Bluetooth functionality expanded to include audio streaming in later models. This allowed drivers to seamlessly connect their smartphones, enabling them to make calls, access their music libraries, and even use voice commands without taking their hands off the wheel, a crucial safety feature in a high-performance vehicle.

As smartphones became ubiquitous, Lamborghini recognized the need for deeper integration. The introduction of the Lamborghini Infotainment System (LIS) in later Gallardo models represented a quantum leap in connectivity. This system offered a more intuitive interface, faster processing, and expanded functionality. It allowed for seamless smartphone integration, providing access to apps, messages, and other phone features directly through the car's interface.

The LIS also introduced features like real-time traffic information, weather updates, and even social media integration. While some purists argued that such connectivity features were unnecessary in a supercar, many customers appreciated the ability to stay connected even while enjoying their high-performance machine.

Compared to its rivals, the Gallardo's infotainment evolution was somewhat conservative. While brands like Ferrari and Porsche were quick to adopt large touchscreens and complex infotainment systems, Lamborghini took a more measured approach. This strategy aligned with the brand's focus on driving purity, ensuring that technology enhanced rather than distracted from the driving experience.

Despite this conservative approach, by the end of its production run, the Gallardo offered an infotainment experience that was competitive with its peers. The system struck a balance between modern connectivity and the raw, driver-focused ethos that Lamborghini is known for.

The evolution of the Gallardo's infotainment and connectivity features reflects broader trends in the automotive industry. It demonstrates how even the most performance-focused vehicles have had to adapt to changing consumer expectations regarding technology and connectivity. The lessons learned from the Gallardo's infotainment journey would go on to inform the development of systems in subsequent Lamborghini models, ensuring that the brand remained relevant in an increasingly connected automotive landscape.

Section 9.7: The Future of Electronics in Lamborghini Supercars

As we look to the horizon of automotive technology, the lessons learned from the Gallardo's electronic evolution continue to shape Lamborghini's approach to innovation. The transition of technology from the Gallardo to its successor, the Huracán, and other models in the Lamborghini lineup, represents a continuous thread of improvement and refinement.

The Gallardo's journey from a relatively analog supercar to a technologically advanced machine provided Lamborghini with invaluable insights. These lessons have been meticulously applied to newer models, ensuring that each successive Lamborghini pushes the boundaries of what's possible in automotive electronics. The Huracán, for instance, debuted with a sophisticated hybrid chassis, combining aluminum and carbon fiber components, controlled by advanced electronic systems that built upon the foundations laid by the Gallardo.

One of the most significant challenges Lamborghini faces is striking the delicate balance between cutting-edge electronics and the raw, visceral driving experience that has long been the hallmark of the brand. The company's approach has been to use technology as an enhancer rather than a replacement for driver engagement. This philosophy is evident in systems like the ANIMA (Adaptive Network Intelligent Management) in the Huracán, which allows drivers to tailor the car's behavior to their preferences, from docile street cruiser to track-focused beast.

Looking ahead, we can anticipate several exciting electronic innovations in future Lamborghini vehicles. The integration of artificial intelligence and machine learning algorithms is likely to play a significant role in optimizing performance and handling characteristics in real-time. We may see advancements in augmented reality displays, providing drivers with immersive, information-rich interfaces that enhance both safety and the driving experience.

Electrification is another frontier where Lamborghini is poised to make significant strides. The lessons learned from the Gallardo's electronic systems will undoubtedly inform the development of hybrid and all-electric powertrains, ensuring that future Lamborghinis maintain their legendary performance while embracing sustainability.

Autonomous driving technologies, while seemingly at odds with Lamborghini's driver-focused ethos, may find their way into future models in subtle ways. For instance, we might see advanced driver assistance systems that can take over in traffic situations but seamlessly hand control back to the driver when the road opens up.

The role of electronics in maintaining Lamborghini's market position is not to be overstated. As competitors continue to push the envelope of automotive technology, Lamborghini must leverage its expertise gained from the Gallardo era to stay at the forefront of innovation. By continuing to develop bespoke electronic systems that enhance rather than dilute the driving experience, Lamborghini can ensure its unique place in the supercar market.

In conclusion, the future of electronics in Lamborghini supercars is bright and full of potential. Building on the legacy of the Gallardo, Lamborghini is well-positioned to continue its tradition of marrying advanced technology with passionate engineering, creating vehicles that are not just fast and beautiful, but also intelligent and forward-thinking. As we move into this exciting future, one thing remains sure: Lamborghini will continue to push the boundaries of what's possible, ensuring that the thrill of driving a raging bull remains unmatched in the automotive world.

Chapter 10: The Gallardos' Impact on Lamborghini's Brand and Market Position

Section 10.1: Redefining Lamborghini's Brand Identity

The introduction of the Gallardo in 2003 marked a significant turning point in Lamborghini's brand identity. For decades, Lamborghini had been synonymous with exclusive, high-performance vehicles produced in limited numbers. The Gallardo, however, ushered in a new era that would redefine what it meant to be a Lamborghini. One of the most striking changes brought about by the Gallardo was the shift from extreme exclusivity to relative accessibility. While still a premium product, the Gallardo was designed to be produced in larger numbers than its predecessors. This move allowed more enthusiasts to experience Lamborghini ownership, broadening the brand's appeal without sacrificing its luxury status. The increased production numbers also helped stabilize Lamborghini's financial position, ensuring the company's long-term viability in a competitive market.

Lamborghini Gallardo: A Decade of Domination

The Gallardo also represented a new balance between raw performance and everyday usability. Previous Lamborghini models were often criticized for being difficult to drive and impractical for regular use. The Gallardo, while still offering breathtaking performance, was engineered to be more user-friendly. Its more compact size, improved visibility, and more forgiving driving dynamics made it a supercar that owners could feasibly use as a daily driver. This shift in philosophy opened up new markets for Lamborghini, attracting customers who wanted supercar thrills without sacrificing practicality. Appealing to a broader audience was a delicate balancing act for Lamborghini.

The challenge lay in attracting new customers without alienating the core enthusiasts who had long been the backbone of the brand. The Gallardo managed this feat by retaining key Lamborghini DNA, its distinctive styling, the naturally aspirated V10 engine, and the visceral driving experience while packaging it in a more approachable form. This strategy allowed Lamborghini to expand its customer base while maintaining its reputation for producing some of the most exciting cars on the road.

The Gallardo effectively became a gateway to the Lamborghini brand. Its (relatively) more attainable price point and easier driving dynamics made it an entry point for those new to the world of supercars. Many first-time Lamborghini owners began with a Gallardo, often progressing to larger, more expensive models like the Murciélago or Aventador as their enthusiasm and budget grew. This created a new customer journey within the Lamborghini brand, fostering long-term loyalty and repeat purchases.

The impact of the Gallardo on Lamborghini's brand perception and desirability cannot be overstated. It transformed Lamborghini from a niche player known for its extreme, sometimes temperamental supercars, into a more rounded, desirable luxury brand. The

Gallardo's success demonstrated that Lamborghini could produce cars that were both thrilling and reliable, both exclusive and (relatively) attainable. This shift in perception increased the brand's overall desirability, attracting a new generation of customers and enthusiasts.

The Gallardos' influence extended beyond sales figures and market share. It rejuvenated Lamborghini's image, making the brand relevant to a wider audience while retaining its aspirational status. The car became a symbol of Lamborghini's ability to evolve and adapt without losing its core identity. This successful rebranding exercise set the stage for Lamborghini's future growth and diversification, paving the way for models such as the Huracán and the Urus SUV.

In essence, the Gallardo redefined what it meant to be a Lamborghini. It proved that the brand could be more than just a producer of extreme, limited-production supercars. Instead, Lamborghini could offer a range of high-performance vehicles that catered to different segments of the luxury market. This redefinition of the brand's identity was crucial in securing Lamborghini's position as a major player in the global luxury automotive market, setting the stage for sustained growth and success in the years to come.

Section 10.2: Market Expansion and Sales Performance

The Lamborghini Gallardo's impact on the company's market expansion and sales performance was nothing short of revolutionary. Over its decade-long production run, the Gallardo achieved unprecedented success, setting new benchmarks for Lamborghini in terms of sales figures, market reach, and overall profitability.

When analyzing the Gallardo's sales figures over its lifetime, the numbers are truly staggering. From its introduction in 2003 to its final year of production in 2013, Lamborghini sold an impressive 14,022 units of the Gallardo. This figure is particularly remarkable when compared to the sales of previous Lamborghini models.

For instance, its predecessor, the Jalpa, saw only 410 units produced over its entire seven-year run from 1981 to 1988. Even the iconic Countach, which was in production for 16 years, managed to sell just 2,049 units. The Gallardo's sales success was unprecedented in Lamborghini's history, accounting for nearly half of all Lamborghinis ever produced since the company's founding in 1963.

The Gallardos' impact on Lamborghini's geographic expansion cannot be overstated. Before its introduction, Lamborghini's market presence was primarily limited to traditional supercar markets such as Europe and North America. However, the Gallardo's relative accessibility and daily usability opened up new markets for the brand.

Emerging economies with growing wealthy classes, such as China, the Middle East, and Southeast Asia, became significant markets for the Gallardo. This geographic diversification not only boosted sales but also helped insulate Lamborghini from regional economic fluctuations.

The financial impact of the Gallardo on Lamborghini's overall revenue and profitability was transformative. While exact figures are closely guarded, industry analysts estimate that the Gallardo accounted for a significant portion of Lamborghini's revenue during its production run. The increased production volume allowed for economies of scale, improving profit margins. Moreover, the Gallardo's success provided Lamborghini with a stable financial foundation, allowing for increased investment in research and development, manufacturing facilities, and future model development.

Lamborghini Gallardo: A Decade of Domination

Perhaps most importantly, the Gallardo played a crucial role in increasing Lamborghini's global market share in the supercar segment. Before the Gallardo, Lamborghini was often seen as a niche player, producing exotic vehicles in limited numbers.

The Gallardo changed this perception, positioning Lamborghini as a serious competitor to established brands like Ferrari and Porsche. By offering a more accessible model without compromising on performance or brand prestige, Lamborghini was able to attract a broader customer base and compete more effectively in the high-performance sports car market.

The Gallardo's success also had a ripple effect on Lamborghini's dealer network. The increased sales volume and broader market appeal justified the expansion of the company's global dealership presence. This expanded network not only supported Gallardo sales but also positioned Lamborghini for future growth with subsequent models.

It's worth noting that the Gallardo's sales performance remained strong throughout its production run, with various special editions and updates helping to maintain interest. This consistent performance demonstrated the model's enduring appeal and helped establish a new baseline for Lamborghini's production and sales expectations. In conclusion, the Gallardos' impact on Lamborghini's market expansion and sales performance was profound and far-reaching.

It transformed Lamborghini from a low-volume, niche manufacturer into a major player in the global supercar market. The model's success provided the financial stability and market presence that would shape Lamborghini's strategy and product development for years to come, setting the stage for future successes, such as the Huracán and Urus.

Section 10.3: Competitive Landscape Shift

The introduction of the Lamborghini Gallardo sent shockwaves through the supercar market, significantly altering the competitive landscape. As Lamborghini's first true volume model, the Gallardo positioned itself as a formidable rival to established players in the high-performance segment, most notably Ferrari and Porsche.

When compared to its key competitors, the Gallardo offered a unique blend of performance, style, and relative accessibility. It went head-to-head with Ferrari's F430 and later the 458 Italia, offering comparable performance but with a distinctly different character. The Gallardo's all-wheel-drive system and more angular design language set it apart from the rear-wheel-drive Ferraris, appealing to a distinct subset of supercar enthusiasts. Against Porsche's 911 Turbo, the Gallardo offered a more exotic and visually striking package, while still providing a level of usability that was previously unheard of in a Lamborghini.

The Gallardos' success forced competitors to rethink their strategies. Ferrari, for instance, responded by expanding its range with models like the California, aimed at a similar market segment that valued both performance and usability. Porsche pushed its 911 range further upmarket and eventually introduced the 918 Spyder hypercar to reassert its performance credentials.

Other manufacturers like McLaren entered the fray with models like the MP4-12C, directly targeting the market space that the Gallardo had proven to be so lucrative.

Perhaps the Gallardo's most significant impact on the competitive landscape was its role in creating and defining the "entry-level" supercar segment. This new category bridged the gap between high-end sports cars and the ultra-exclusive hypercars, offering supercar performance and prestige at a relatively more attainable price point. The success of this segment led to a proliferation of models from various manufacturers, all vying for a piece of this newly expanded market.

The introduction of the Gallardo also had a profound effect on pricing strategies within the supercar market. By offering a more "affordable" option, Lamborghini forced competitors to reconsider their pricing structures. This led to a more stratified market, with more precise distinctions between entry-level supercars, flagship models, and limited-edition hypercars. The Gallardo's pricing strategy made ownership of a true supercar a reality for a broader range of enthusiasts, effectively democratizing supercar ownership to an extent never seen before.

This democratization of supercar ownership was perhaps the Gallardo's most transformative impact on the competitive landscape. By making Lamborghini ownership more accessible, it expanded the overall market for high-performance vehicles. This not only benefited Lamborghini but also created opportunities for other manufacturers to enter the market or expand their offerings in this segment.

The Gallardos' influence extended beyond just sales figures and market share. It changed perceptions of what a supercar could be, challenging the notion that these vehicles were only for weekend drives or track days. By proving that a supercar could be both thrilling and relatively practical, the Gallardo shifted consumer expectations and forced competitors to focus more on usability and everyday performance alongside raw speed and power.

In essence, the Lamborghini Gallardo didn't just compete in the supercar market; it fundamentally reshaped it. Its success story rewrote the rules of engagement in this high-stakes segment, influencing everything from product development and pricing strategies to brand positioning and market expansion. The ripple effects of the Gallardo's impact on the competitive landscape continue to be felt in the supercar market to this day, cementing its place as one of the most influential models in automotive history.

Section 10.4: Technological Advancements and Innovation

The Lamborghini Gallardo was not only a commercial success but also a technological tour de force that showcased the company's engineering prowess and innovative spirit. This section delves into how the Gallardo became a platform for Lamborghini's technological advancements and how it influenced the company's approach to innovation.

The Gallardo served as a technological showcase for Lamborghini, introducing several cutting-edge features that would later become hallmarks of the brand. One of the most significant innovations was the integration of an all-wheel-drive system, a first for a V10-powered supercar. This system, derived from Audi's Quattro technology, provided exceptional traction and handling, allowing drivers to harness the car's immense power with greater confidence. The Gallardo's aluminum space frame chassis was another technological marvel, offering a perfect balance of rigidity and lightweight construction.

The success of the Gallardo had a profound impact on Lamborghini's research and development focus. The increased sales and revenue generated by the model allowed the company to invest more heavily in R&D, pushing the boundaries of what was possible in supercar design and engineering. This investment paid dividends not only for the Gallardo's continuous improvement over its lifetime but also for the development of future models.

Lamborghini Gallardo: A Decade of Domination

The technological advancements pioneered in the Gallardo had a significant trickle-down effect on other Lamborghini models. Features such as the advanced all-wheel-drive system, lightweight construction techniques, and high-performance engine management systems were refined in the Gallardo and later implemented across the Lamborghini range. This approach enabled the company to amortize development costs across multiple models, ensuring that even its most exclusive vehicles benefited from proven technologies.

The Gallardo's development also fostered increased collaboration and knowledge sharing within the Volkswagen Group, Lamborghini's parent company. The synergies between Lamborghini, Audi, and other group brands allowed for the exchange of expertise in areas such as all-wheel-drive systems, lightweight materials, and engine technology. This collaboration not only benefited Lamborghini but also contributed to technological advancements across the entire Volkswagen Group.

Perhaps most importantly, the Gallardo served as a testbed for future Lamborghini innovations. Its long production run allowed the company to continuously refine and improve various technologies, from engine management systems to aerodynamics. The lessons learned from the Gallardo's development and evolution directly influenced the design and engineering of its successor, the Huracán, as well as other models, such as the Aventador.

The Gallardos' impact on Lamborghini's technological trajectory cannot be overstated. It pushed the company to innovate in areas such as materials science, powertrain technology, and vehicle dynamics. The car's success also justified increased investment in advanced manufacturing techniques and state-of-the-art production facilities, setting the stage for Lamborghini's future as a high-tech supercar manufacturer.

Moreover, the Gallardo's technological achievements helped to cement Lamborghini's reputation as a serious player in the supercar market, capable of producing vehicles that were not only visually striking and emotionally stirring but also technologically advanced and engineered to the highest standards. This perception shift was crucial in attracting a new generation of customers who demanded cutting-edge technology alongside traditional supercar attributes.

In conclusion, the Gallardo was much more than just another model in Lamborghini's lineup; it was a technological pioneer that drove innovation throughout the company and beyond. Its impact on Lamborghini's approach to R&D, its role in fostering collaboration within the Volkswagen Group, and its function as a testbed for future innovations all contributed to shaping Lamborghini into the technologically advanced supercar manufacturer it is today. The technological legacy of the Gallardo continues to influence Lamborghini's product development strategy, ensuring that the brand remains at the forefront of automotive innovation.

Section 10.5: Brand Partnerships and Cultural Impact

The Lamborghini Gallardo's influence extended far beyond the automotive world, making a significant impact on popular culture and reshaping Lamborghini's approach to brand partnerships. From the silver screen to the digital realm, the Gallardo became a ubiquitous symbol of automotive excellence and aspiration.

In Hollywood, the Gallardo quickly became a favorite among filmmakers looking to add a touch of exotic flair to their productions. Its sleek lines and unmistakable engine note graced numerous blockbusters, cementing its status as a cultural icon. The car's appearance in films like "The Dark Knight" and "Iron Man" not only showcased its visual appeal but also associated the Lamborghini brand with heroism and cutting-edge technology.

Lamborghini Gallardo: A Decade of Domination

The gaming industry also embraced the Gallardo, featuring it prominently in popular racing franchises such as "Need for Speed," "Forza Motorsport," and "Gran Turismo." This digital presence allowed millions of gamers worldwide to experience the thrill of driving a Lamborghini, albeit virtually. The Gallardo's inclusion in these games served as a powerful marketing tool, introducing the brand to younger audiences and fostering a new generation of Lamborghini enthusiasts.

Recognizing the Gallardo's broad appeal, Lamborghini strategically pursued collaborations with luxury brands outside the automotive sector. Limited edition models were created in partnership with high-end watchmakers, fashion houses, and technology companies. These collaborations not only resulted in unique, highly sought-after versions of the Gallardo but also expanded Lamborghini's reach into new markets and demographics.

The Gallardo's cultural impact profoundly influenced Lamborghini's marketing and promotional strategies. The company shifted from traditional automotive advertising to a more lifestyle-oriented approach, positioning the Gallardo as not just a car, but a symbol of success, passion, and individuality. This strategy involved increasing presence at high-profile events, sponsoring luxury lifestyle experiences, and forming partnerships with influential personalities across various industries.

In the realm of automotive enthusiasm and car culture, the Gallardo played a pivotal role in democratizing the supercar experience. Its relative accessibility compared to previous Lamborghini models meant that more enthusiasts could realistically aspire to ownership. This led to the growth of Lamborghini owner clubs, track day events, and online communities dedicated to the Gallardo, fostering a sense of camaraderie among owners and admirers alike.

Perhaps most significantly, the Gallardo's cultural presence had a lasting impact on younger generations of car enthusiasts. For many millennials and Gen Z individuals, the Gallardo was the poster car of their youth, adorning bedroom walls and computer screens. This early exposure to the brand created a strong emotional connection, influencing future purchasing decisions and career aspirations within the automotive industry.

The Gallardo's cultural impact went beyond mere visibility; it reshaped perceptions of what a supercar could be. By balancing exotic appeal with relative attainability, it challenged the notion that supercars were solely the domain of the ultra-wealthy. This shift in perception opened doors for Lamborghini to engage with a broader audience, setting the stage for future models to continue this legacy of accessible performance and style.

In essence, the Gallardo became more than just a car; it evolved into a cultural phenomenon that transcended the automotive world. Its presence in popular media, strategic brand partnerships, and influence on car culture not only elevated Lamborghini's brand image but also played a crucial role in shaping the desires and aspirations of an entire generation of car enthusiasts. The Gallardo's cultural impact ensured that Lamborghini remained not just relevant but deeply ingrained in the zeitgeist of the early 21st century.

Section 10.6: Production and Manufacturing Evolution

The Lamborghini Gallardo not only revolutionized the company's market position but also had a profound impact on its production and manufacturing processes. As Lamborghini's first true volume model, the Gallardo necessitated a significant shift in the company's approach to vehicle production.

Lamborghini Gallardo: A Decade of Domination

Before the Gallardo, Lamborghini was known for its low-volume, highly exclusive supercars. The introduction of the Gallardo required the company to scale up its production capabilities dramatically. This transition posed a unique challenge: how to increase production volume while maintaining the high quality and craftsmanship associated with the Lamborghini brand.

To meet this challenge, Lamborghini invested heavily in modernizing its manufacturing facilities. The Sant'Agata Bolognese factory underwent extensive renovations and expansions to accommodate the increased production demands. New assembly lines were introduced, incorporating advanced robotics and automation to enhance efficiency without compromising quality. This modernization effort not only enabled Lamborghini to produce the Gallardo in higher numbers but also set the stage for future models.

The increased production volume of the Gallardo had a significant impact on Lamborghini's supply chain.

The company had to forge new partnerships with suppliers capable of meeting the higher demand for components. This shift required careful management to ensure that each supplier could maintain the exacting standards necessary for a Lamborghini. The experience gained in managing this expanded supply chain proved invaluable for future models and contributed to the company's overall growth strategy.

Quality control became more critical than ever with the increased production volume. Lamborghini implemented rigorous quality assurance processes throughout the manufacturing cycle. Each Gallardo underwent extensive testing and inspection before leaving the factory, ensuring that the higher production numbers did not come at the cost of Lamborghini's renowned quality.

Lamborghini Gallardo: A Decade of Domination

The Gallardo's production run also saw Lamborghini significantly expand its workforce. The company hired and trained numerous skilled workers, from engineers to artisans, to support the increased production. This expansion of human resources not only facilitated Gallardo's production but also laid the foundation for a pool of expertise that would benefit future Lamborghini models.

One of the most important aspects of the Gallardo's influence on Lamborghini's manufacturing was the lessons learned throughout its production run. The company gained valuable insights into efficient production methods, quality control for higher volumes, and the management of a more complex supply chain. These lessons were instrumental in shaping Lamborghini's approach to future model production strategies.

The manufacturing evolution driven by the Gallardo also had a positive impact on Lamborghini's ability to introduce variants and special editions. The more flexible production setup allowed for more straightforward implementation of design changes and limited-run models, enhancing the brand's ability to cater to different market segments and maintain excitement throughout the Gallardo's long production life.

By the end of the Gallardo's production run, Lamborghini had transformed from a boutique manufacturer into a company capable of producing thousands of high-quality supercars annually. This evolution in production and manufacturing capabilities played a crucial role in Lamborghini's overall success and market growth during the Gallardo era.

The experience and infrastructure developed during the Gallardo's production laid the groundwork for the successful launch of its successor. They even influenced the company's ability to diversify into new segments with models like the Urus SUV. The

Gallardo thus not only changed Lamborghini's market position but also fundamentally transformed how the company approached the art and science of supercar manufacturing.

Section 10.7: Long-term Strategic Implications

The Lamborghini Gallardo's impact extended far beyond its impressive sales figures and immediate market success. This groundbreaking model laid the foundation for Lamborghini's long-term strategic vision, influencing the company's approach to product development, market positioning, and overall business strategy for years to come.

The Gallardos's influence on Lamborghini's long-term product strategy cannot be overstated. Its success in balancing high performance with relative accessibility set a new benchmark for future models. This approach led to a shift in Lamborghini's product development philosophy, emphasizing the creation of vehicles that could appeal to a broader audience while still maintaining the brand's core values of extreme performance and striking design.

Perhaps the most direct result of the Gallardo's success was its impact on the development of its successor, the Huracán. Lamborghini took the lessons learned from the Gallardo's decade-long production run and applied them to create a car that would build upon its predecessor's strengths while addressing its weaknesses.

The Huracán's design, performance, and market positioning were all heavily influenced by the Gallardo's legacy, resulting in a car that successfully carried forward the torch lit by its predecessor. The Gallardo also played a crucial role in shaping Lamborghini's approach to model lifecycles. Before the Gallardo, Lamborghini's models typically had shorter production runs with fewer variations.

Lamborghini Gallardo: A Decade of Domination

The Gallardo's long production life, coupled with numerous special editions and updates, demonstrated the value of evolving a successful platform over time. This strategy allowed Lamborghini to amortize development costs over a longer period while keeping the model fresh and relevant in the market.

Interestingly, the Gallardo's success also influenced Lamborghini's diversification strategy. By proving that the brand could successfully expand its appeal without diluting its core values, the Gallardo paved the way for more radical departures from tradition. The most notable example of this is the introduction of the Urus SUV. While an SUV might have seemed incongruous with Lamborghini's supercar image in the pre-Gallardo era, the brand's expanded appeal and market presence made such a move not only possible but highly successful.

The Gallardo's legacy in shaping Lamborghini's future vision cannot be underestimated. It showed that Lamborghini could be more than just a niche player producing exotic supercars in limited numbers. The Gallardo proved that with the right product and strategy, Lamborghini could become a significant force in the high-performance automotive market.

This shift in perspective has influenced every aspect of Lamborghini's operations, from research and development to marketing and sales strategies. The company now approaches each new model with the intention of balancing performance, design, and accessibility, always seeking to push the boundaries of what's possible while ensuring that its vehicles remain attainable to a broader range of enthusiasts.

Moreover, the Gallardo's success has given Lamborghini the confidence and resources to invest heavily in future technologies. From hybrid powertrains to advanced materials, Lamborghini is now at the forefront of automotive innovation, a position that owes much to the financial and brand strength built during the Gallardo era.

In essence, the Gallardo didn't just change Lamborghini's present; it reshaped its future. It transformed Lamborghini from a company known for producing beautiful but rare supercars into a technological powerhouse and a major player in the global luxury automotive market. As Lamborghini continues to evolve and grow, the strategic lessons learned from the Gallardo's success continue to guide its path forward, ensuring that the brand remains at the cutting edge of performance, design, and innovation in the automotive world.

Lamborghini Gallardo: A Decade of Domination

Chapter 11: Comparing Generations: How the Gallardo Evolved Over a Decade

Section 11.1: First Generation Gallardo (2003-2008)

The dawn of a new era for Lamborghini began in 2003 with the introduction of the Gallardo. This groundbreaking model marked a significant departure from the company's traditional offerings, positioning itself as a more accessible yet equally thrilling entry point into the world of Lamborghini.

The first Gallardo rolled off the production line in 2003, boasting a 5.0-liter V10 engine producing 493 horsepower. This powerplant was a testament to Lamborghini's engineering prowess, delivering an intoxicating blend of performance and aural drama. The engine's positioning, mounted longitudinally behind the cabin, not only contributed to the car's excellent weight distribution but also became a signature element of the Gallardo's design.

Lamborghini Gallardo: A Decade of Domination

Performance was, unsurprisingly, a key focus for the original Gallardo. With a 0-60 mph time of 4.2 seconds, it set a new standard for Lamborghini's entry-level supercar. The top speed of 192 mph ensured that the Gallardo could hold its own against more expensive rivals. However, it's just about straight-line speed; the Gallardo's all-wheel-drive system and finely-tuned suspension provided a level of handling precision that was previously unheard of in a Lamborghini of this class.

The first-generation Gallardo's sleek, wedge-shaped design was a departure from Lamborghini's traditionally angular aesthetic. Designer Luc Donckerwolke crafted a shape that was at once unmistakably Lamborghini yet more understated than its larger sibling, the Murciélago. The compact dimensions and clean lines gave the Gallardo a sense of agility and purpose, while distinctive details like the canted headlights and large air intakes ensured it commanded attention on the road.

Throughout its initial production run, the Gallardo lineup expanded to include several variants and special editions. The introduction of the Gallardo Spyder in 2005 brought open-top thrills to the range, its electrohydraulic soft top allowing drivers to immerse themselves in the V10's symphonic exhaust note fully. The Gallardo SE (Special Edition) and Nera special editions offered unique aesthetic packages. At the same time, the Superleggera variant, introduced in 2007, pushed the performance envelope even further with its lightweight construction and power bump to 523 horsepower.

The reception of the first-generation Gallardo was overwhelmingly positive. Critics praised its blend of everyday usability and supercar performance, while customers flocked to Lamborghini showrooms in unprecedented numbers. The Gallardo quickly became Lamborghini's best-selling model, with over 7,000 units sold by 2008. This success not only solidified the Gallardo's place in Lamborghini's lineup but also played a crucial role in the brand's financial resurgence under Audi ownership.

The impact of the first-generation Gallardo on the supercar market is undeniable. It forced competitors to reevaluate their offerings in the segment and set a new benchmark for what customers could expect from an 'entry-level' supercar. The Gallardo's success also paved the way for future V10-powered Lamborghinis, establishing a lineage that continues to this day.

As the first chapter in the Gallardo's decade-long story, the 2003-2008 models laid a robust foundation. They proved that Lamborghini could produce a more accessible supercar without diluting the brand's core values of passion, performance, and visual drama. This generation of the Gallardo not only met the high expectations set for it but exceeded them, setting the stage for even more impressive iterations to come.

Section 11.2: Gallardo LP560-4 (2008-2013)

In 2008, Lamborghini unveiled a significant update to the Gallardo lineup with the introduction of the LP560-4. This refreshed model marked a substantial leap forward in performance, design, and technology, cementing the Gallardo's position as a dominant force in the supercar market.

The heart of the LP560-4 was its new powerplant, a larger 5.2-liter V10 engine that produced an impressive 552 horsepower. This substantial increase in power, combined with other mechanical refinements, resulted in a dramatic improvement in performance. The LP560-4 could now accelerate from 0 to 60 mph in just 3.7 seconds, shaving 0.3 seconds off its predecessor's time. This enhanced performance wasn't just about straight-line speed; the LP560-4 also boasted a higher top speed and improved overall driving dynamics.

Visually, the LP560-4 received a comprehensive styling update that gave it a more aggressive and contemporary appearance. The revised front fascia featured larger air intakes and new headlights, which not only enhanced the car's visual appeal but also improved cooling efficiency and aerodynamics. The rear end also saw significant changes, with new taillights and a redesigned diffuser that contributed to the car's enhanced high-speed stability.

Under the skin, the LP560-4 benefited from numerous technological advancements. One of the most notable was the updated e-gear transmission, which offered faster shift times and a new 'Corsa' mode explicitly designed for track use. This new transmission not only improved performance but also enhanced the overall driving experience, providing smoother shifts during everyday driving and lightning-fast gear changes when pushing the car to its limits.

The LP560-4 also saw improvements in its all-wheel-drive system, which now featured a more rear-biased torque split for sportier handling characteristics. Combined with revised suspension tuning, these changes resulted in a car that was more agile and responsive than its predecessor, yet still maintained the all-weather capability that had become a Gallardo hallmark.

Throughout the LP560-4's production run, Lamborghini introduced several variants and special editions that further expanded the Gallardo's appeal. One of the most notable was the LP570-4 Superleggera, introduced in 2010. This track-focused version took the Gallardo's performance to new heights with its lightweight construction, shedding over 70 kg compared to the standard LP560-4. The Superleggera also received a slight power bump to 570 horsepower, resulting in even more blistering performance.

Lamborghini Gallardo: A Decade of Domination

Other notable variants included the LP550-2, which ditched the all-wheel-drive system in favor of a purist-pleasing rear-wheel-drive layout, and the LP560-4 Spyder, which combined the coupe's performance upgrades with the thrill of open-top driving.

The introduction of the LP560-4 and its subsequent variants marked a significant evolution for the Gallardo. It addressed many of the criticisms of the original model while building upon its strengths. The improved performance, refined styling, and advanced technology allowed the Gallardo to remain competitive in an increasingly crowded supercar market.

Moreover, the LP560-4 demonstrated Lamborghini's commitment to continuous improvement and innovation. By significantly updating the Gallardo midway through its life cycle, Lamborghini showed that it was not content to rest on its laurels. This philosophy of constant evolution would become a hallmark of the brand, influencing the development of future models.

The LP560-4's success in the market was evident, with sales remaining strong throughout its production run. It continued to attract new buyers to the brand while also appealing to existing Lamborghini enthusiasts. The model's balance of extreme performance, everyday usability, and unmistakable Italian flair ensured its place as one of the most desirable supercars of its era.

In retrospect, the LP560-4 represents a crucial chapter in the Gallardo's story. It took an already successful formula and refined it, creating a car that was faster, more capable, and more technologically advanced than its predecessor. This evolution set the stage for the final years of Gallardo production, during which Lamborghini would continue to push the boundaries of what was possible with its "entry-level" supercar.

Section 11.3: Final Updates and Special Editions (2012-2013)

As the Lamborghini Gallardo approached the end of its remarkable production run, the Italian automaker gave its beloved supercar one last hurrah. In 2012, Lamborghini unveiled the final facelift for the Gallardo, featuring sharper styling cues inspired by its bigger brother, the Aventador. This update breathed new life into the decade-old design, ensuring the Gallardo would leave the stage as fresh and desirable as ever.

While the power output of the 5.2-liter V10 engine remained unchanged at 552 horsepower for the standard models, Lamborghini's engineers focused on refining the Gallardo's handling characteristics. The suspension received careful tuning, resulting in improved cornering stability and a more responsive driving experience. These subtle yet effective changes allowed the Gallardo to maintain its competitive edge in an increasingly crowded supercar market.

Looking back at the Gallardo's design evolution from 2003 to 2013, one can't help but marvel at its transformation. The original model, with its smooth, understated lines, represented a departure from Lamborghini's traditionally angular aesthetic. Over the years, each iteration became progressively more aggressive and angular, culminating in the final version's razor-sharp appearance. This evolution not only reflected changing design trends but also Lamborghini's shift towards a more track-focused philosophy.

The twilight of the Gallardo's production run saw the introduction of several limited-edition models, each more exclusive than the last. The most notable of these was the Gallardo LP 560-4 Spyder Finale, limited to just 50 units. This special edition served as a fitting tribute to the model that had become synonymous with Lamborghini's 21st-century renaissance. Featuring unique color schemes and bespoke interior touches, these final editions became instant collectors' items, sought after by enthusiasts and investors alike.

As production wound down in 2013, the Gallardo's impact on Lamborghini and the broader supercar market became clear. With over 14,000 units sold throughout its lifetime, the Gallardo had not only become the most successful Lamborghini model in history but had also redefined what it meant to be an "entry-level" supercar. Its accessibility (relative to other exotic marques) brought new customers to the brand, expanding Lamborghini's global presence and financial stability.

The Gallardo's legacy extends far beyond mere sales figures, however. It proved that Lamborghini could produce a reliable, everyday-drivable supercar without sacrificing the drama and excitement expected of the brand. This successful formula laid the groundwork for its successor, the Huracán, and influenced the development of other models in the Lamborghini lineup.

Moreover, the Gallardo's evolution over its production run demonstrated Lamborghini's commitment to continuous improvement and innovation. From its original incarnation to the final special editions, each update and refinement pushed the boundaries of performance, technology, and design. This relentless pursuit of excellence not only kept the Gallardo relevant in a fast-moving market but also set new standards for the entire supercar industry.

As the last Gallardo rolled off the production line in 2013, it marked the end of an era for Lamborghini. Yet, it also represented a new beginning. The lessons learned, technologies developed, and customer relationships forged during the Gallardo's decade-long run would go on to shape Lamborghini's future, ensuring that the spirit of this revolutionary supercar would live on in the next generation of Raging Bulls.

Section 11.4: Performance Comparison Across Generations

Significant performance improvements across all generations marked the Lamborghini Gallardo's evolution over its decade-long production run. This section explores the key areas where the Gallardo underwent substantial enhancements, highlighting how Lamborghini's relentless pursuit of perfection transformed its entry-level supercar into a formidable performance machine.

At the heart of the Gallardo's performance evolution was its engine. The original 2003 model debuted with a 5.0-liter V10 engine, producing a respectable 493 horsepower. However, as the years progressed, Lamborghini's engineers continually refined and improved this powerplant.

The introduction of the LP560-4 in 2008 saw a significant leap forward, with the engine capacity increased to 5.2 liters and power output boosted to 552 horsepower. By the end of its production run, the most potent Gallardo variants were churning out an impressive 570 horsepower. This represents a power increase of nearly 20% over the course of a decade, a testament to Lamborghini's commitment to pushing the boundaries of performance.

The improvements in engine output directly translated to enhanced acceleration and top speed figures. The original Gallardo could sprint from 0-60 mph in a brisk 4.2 seconds, a figure that was highly impressive for its time. However, as the Gallardo evolved, this number steadily decreased. The LP560-4 shaved this down to 3.7 seconds, while the final Superleggera models could achieve the same feat in a blistering 3.4 seconds. Top speeds also saw a notable increase, rising from the original model's 192 mph to over 200 mph in later variants. These improvements in straight-line performance ensured that the Gallardo remained competitive throughout its lifespan, even as rival manufacturers introduced newer, more advanced models.

While raw speed is impressive, a true supercar must also excel in handling and driving dynamics. In this regard, the Gallardo saw continuous refinement throughout its production run. The all-wheel-drive system, a hallmark feature of the Gallardo, underwent several iterations. Later models benefited from more sophisticated AWD systems that could more effectively distribute power between the front and rear axles, resulting in improved traction and more neutral handling characteristics.

Suspension tuning also saw ongoing improvements, with later models offering sharper turn-in, reduced body roll, and better overall balance. The introduction of magnetorheological dampers in some variants further enhanced the Gallardo's ability to switch between comfortable road manners and track-focused precision.

Aerodynamics played a crucial role in the Gallardo's performance evolution. The introduction of the LP560-4 in 2008 marked a significant leap forward in this area, with Lamborghini claiming a 31% improvement in aerodynamic efficiency compared to the original model.

This was achieved through a combination of subtle styling changes and the implementation of more advanced underbody aerodynamics. The result was a car that not only cut through the air more efficiently but also generated more downforce, enhancing high-speed stability and cornering performance. Subsequent models, particularly the track-focused Superleggera variants, further built upon these improvements with additional aerodynamic elements such as larger rear diffusers and more aggressive front splitters.

Weight reduction was another key focus area in the Gallardo's evolution. While the basic structure of the car remains essentially unchanged throughout its production run, Lamborghini made concerted efforts to shed weight wherever possible. This was most evident in the Superleggera variants, which showcased Lamborghini's expertise in carbon fiber construction.

The final Gallardo Superleggera models weighed up to 100 kg less than their standard counterparts, thanks to extensive use of carbon fiber for body panels, interior components, and even engine bay parts. This weight reduction not only improved acceleration and handling but also enhanced braking performance and overall efficiency.

The cumulative effect of these performance enhancements transformed the Gallardo from an impressive but relatively docile supercar into an actual track weapon. By the end of its production run, the Gallardo was setting lap times that would have been unthinkable for the original model.

More importantly, these improvements were not just about raw numbers but also about the overall driving experience. Later Gallardo models were praised for their more engaging and visceral character, offering a level of driver involvement that matched their impressive performance figures.

In conclusion, the Gallardo's performance evolution over its decade-long production run was nothing short of remarkable. From its engine and acceleration to its handling and aerodynamics, every aspect of the car saw significant improvements. This continual refinement not only kept the Gallardo competitive in an increasingly crowded supercar market but also set new benchmarks for performance and driving excitement. The lessons learned and technologies developed during the Gallardo's evolution would go on to inform future Lamborghini models, ensuring that its legacy would extend far beyond its production run.

Section 11.5: Technological Evolution

The Lamborghini Gallardo's decade-long production run witnessed significant advancements in automotive technology, and the iconic supercar evolved to incorporate these innovations. This section explores the key technological developments that shaped the Gallardo's progression from 2003 to 2013.

One of the most notable areas of improvement was in the Gallardo's transmission systems. The early models offered a traditional six-speed manual gearbox and an optional "e-gear" automated manual transmission. As the years progressed, Lamborghini continually refined the e-gear system, dramatically reducing shift times.

The original e-gear transmission completed gear changes in around 120 milliseconds, but by the end of the Gallardo's production run, this had been slashed to a mere 50 milliseconds. These improvements not only enhanced performance but also provided a smoother, more responsive driving experience.

The Gallardo also saw significant advancements in electronic aids and driver assistance features. Early models were relatively simple by modern standards, but later iterations introduced sophisticated traction control systems, launch control, and more advanced stability control programs. These systems allowed drivers to push the car closer to its limits while maintaining a higher degree of safety. The introduction of the "ANIMA" (Adaptive Network Intelligent Management) system in later models allowed drivers to select different driving modes, adjusting the car's behavior to suit various conditions and preferences.

Infotainment and connectivity features in the Gallardo evolved significantly over its lifespan. The earliest models featured basic audio systems, reflecting the car's focus on pure driving experience. However, as customer expectations changed, later Gallardos began to offer more advanced features.

By the end of its run, the Gallardo was available with sophisticated navigation systems, multimedia interfaces, and improved connectivity options. While never losing sight of its performance-focused ethos, these additions made the Gallardo more versatile and appealing to a broader range of buyers.

Material science and construction techniques saw considerable advancement during the Gallardo's production. Lamborghini increasingly employed carbon fiber and other lightweight materials in the car's construction, particularly in special editions like the Superleggera.

These advancements allowed for significant weight reductions without compromising structural integrity. For instance, the final Gallardo LP 570-4 Superleggera weighed approximately 100 kg less than the standard model, thanks to extensive use of carbon fiber components.

Safety features also saw notable improvements over the Gallardo's lifespan. While early models met all contemporary safety standards, later versions benefited from advancements in crash protection technology. Improved crumple zones, more sophisticated airbag systems, and enhanced structural rigidity all contributed to making the later Gallardos safer than their predecessors. Additionally, the aforementioned electronic stability systems played a dual role in improving both performance and safety.

The evolution of the Gallardo's braking system is another area worthy of note. While the car always featured impressive stopping power, later models saw the introduction of carbon-ceramic brakes as an option. These offered reduced weight, improved heat dissipation, and enhanced durability compared to traditional steel brakes, further improving the Gallardo's overall performance envelope.

Aerodynamics was an area of continuous refinement throughout the Gallardo's production. The introduction of the LP560-4 in 2008 brought a significant leap in aerodynamic efficiency, with Lamborghini claiming a 31% improvement over the original model. This was achieved through subtle but effective changes to the body shape, underbody design, and the introduction of more sophisticated airflow management systems.

The Gallardo's engine management systems also saw continuous improvement. Later models featured more advanced engine control units, allowing for better power delivery, improved fuel efficiency, and reduced emissions. This technological progression enabled Lamborghini to extract more power from the V10 engine while meeting increasingly stringent environmental regulations.

In conclusion, the technological evolution of the Lamborghini Gallardo over its ten-year production run was nothing short of remarkable. From transmission refinements and advanced driver aids to improvements in materials and safety features, each iteration of the Gallardo showcased Lamborghini's commitment to innovation. These advancements not only enhanced the car's performance and drivability but also improved its everyday usability and appeal to a broader audience. The Gallardo's technological journey set the stage for future Lamborghini models, demonstrating the brand's ability to blend traditional supercar attributes with cutting-edge automotive technology.

Section 11.6: Design Language Evolution

The Lamborghini Gallardo's design language underwent a remarkable transformation throughout its decade-long production run, reflecting not only advancements in automotive design but also Lamborghini's evolving brand identity. This evolution is evident in both the exterior and interior design, as well as in the expanding range of customization options offered to discerning customers.

The exterior styling of the Gallardo saw the most dramatic changes over the years. The original 2003 model featured a relatively understated design for a Lamborghini, with smooth, flowing lines and rounded edges. This initial design was a departure from the angular, aggressive styling of previous Lamborghini models, aiming to appeal to a broader audience. However, as the Gallardo matured, its exterior became increasingly sharp and angular.

The 2008 introduction of the LP560-4 marked a significant shift in the Gallardo's design language. The front fascia became more aggressive, with larger air intakes and a more pronounced front splitter. The headlights, once rounded, took on a more angular shape, giving the car a more menacing appearance. The side profile saw the introduction of sharper creases and more defined character lines, while the rear end received redesigned taillights and a more prominent diffuser.

This trend towards a more aggressive aesthetic continued with subsequent updates and special editions. The 2012 facelift brought the Gallardo's design more in line with its larger sibling, the Aventador, featuring even sharper lines and more pronounced aerodynamic elements. This evolution from soft, rounded forms to sharp, angular surfaces not only gave the Gallardo a more dynamic appearance but also improved its aerodynamic efficiency.

The interior design of the Gallardo also saw significant changes over the years, albeit more subtle than the exterior transformations. Early models featured a relatively simple cabin design, with a focus on driver-centric controls and high-quality materials. As the model evolved, the interior became more refined and technologically advanced.

Later Gallardo models featured more sculpted seats, providing better support during high-performance driving. The dashboard and center console saw the introduction of more advanced infotainment systems, with larger screens and improved functionality. The quality of materials also saw an upgrade, with increased use of fine leather, Alcantara, and carbon fiber trim options.

One of the most significant developments in the Gallardo's design evolution was the expansion of customization options. The introduction of Lamborghini's Ad Personam program midway through the Gallardo's production run allowed customers unprecedented levels of personalization. This program offered an extensive range of exterior paint colors, including matte finishes and multi-layer pearl effects. Interior customization options expanded to include a wide variety of leather and Alcantara colors, as well as contrasting stitching and piping.

Special edition models played a crucial role in the Gallardo's design evolution, often introducing unique design elements that would later influence the broader Lamborghini lineup. The Gallardo Bicolore, for instance, showcased a striking two-tone paint scheme that highlighted the car's angular lines and would go on to inspire similar treatments in future Lamborghini models. The Superleggera editions, with their extensive use of carbon fiber and track-focused aesthetics, further pushed the boundaries of the Gallardo's design language.

The evolution of the Gallardo's design reflects broader changes in Lamborghini's overall design philosophy. As the brand shifted towards a more aggressive, track-focused aesthetic, the Gallardo followed suit. This shift not only affected the Gallardo but also set the stage for future Lamborghini models, including its successor, the Huracán.

In conclusion, the design language evolution of the Lamborghini Gallardo over its ten-year production run was nothing short of remarkable. From its initially understated appearance to its final, aggressively styled iterations, the Gallardo's design transformation mirrored Lamborghini's journey as a brand. This evolution not only kept the model fresh and exciting throughout its lifespan but also played a crucial role in shaping the future of Lamborghini's design language, leaving an indelible mark on the automotive design landscape.

Section 11.7: Market Position and Competitor Analysis

The Lamborghini Gallardo's decade-long production run was characterized by an impressive sales performance and a continually evolving market position. Throughout its lifespan, the Gallardo consistently outperformed sales expectations, cementing its place as Lamborghini's best-selling model of all time. In its peak year of 2008, coinciding with the introduction of the LP560-4, Lamborghini sold over 2,000 Gallardo units, a testament to the model's enduring appeal and the success of its mid-cycle refresh.

Despite the continuous improvements and updates made to the Gallardo over the years, Lamborghini managed to maintain a relatively stable pricing strategy. The base price of the Gallardo saw only modest increases throughout its production run, with adjustments primarily accounting for inflation and added features. This pricing discipline helped the Gallardo maintain its position as an "entry-level" supercar, making it more accessible to a broader range of enthusiasts and collectors.

When compared to its key competitors, such as the Ferrari F430 and the Porsche 911 Turbo, the Gallardo held its own admirably. While the Ferrari F430 may have offered a more exotic V8 soundtrack and the cache of the prancing horse badge, the Gallardo countered with superior all-weather performance thanks to its advanced all-wheel-drive system. The Porsche 911 Turbo, known for its everyday usability, found a worthy adversary in the Gallardo, which combined exotic Italian styling with surprising practicality for a mid-engined supercar.

The Gallardo's success had a profound impact on Lamborghini's brand image. Prior to the Gallardo, Lamborghini was primarily known as a low-volume manufacturer of exotic, somewhat temperamental supercars. The Gallardo's reliability, relative affordability, and increased production volumes helped transform Lamborghini into a more mainstream luxury sports car brand. This shift allowed Lamborghini to compete more directly with established players like Ferrari and Porsche, significantly expanding its market presence and global recognition.

Perhaps most importantly, the Gallardo's evolution over its ten-year run heavily influenced the development of future Lamborghini models. Many of the technological advancements, performance improvements, and design innovations refined during the Gallardo's lifespan were directly applied to its successor, the Huracán. The lessons learned from producing a higher-volume model also informed Lamborghini's overall production strategies and quality control processes, benefiting the entire model range.

The Gallardo's market performance also demonstrated the viability of a more accessible Lamborghini model, paving the way for the brand to explore new market segments. This success likely played a role in Lamborghini's decision to develop the Urus SUV, further broadening the brand's appeal and market reach.

In the context of the broader automotive industry, the Gallardo's enduring success and constant evolution set new benchmarks for the supercar segment. It proved that a supercar could be both exotic and relatively practical, high-performance and reasonably reliable. This formula not only challenged established competitors but also inspired new entrants to the supercar market, elevating the entire segment in terms of performance, technology, and everyday usability.

As the Gallardo's production run came to a close in 2013, it left behind a legacy of innovation, performance, and market success that would continue to influence Lamborghini and the broader supercar industry for years to come. The model's ability to remain relevant and desirable over a decade-long production run is a testament to its fundamental excellence and Lamborghini's skill in evolving the platform to meet changing market demands and technological possibilities.

Chapter 12: Legacy and Influence: The Gallardos' Place in Supercar History

Section 12.1: The Gallardos' Impact on Lamborghini

The Lamborghini Gallardo's influence on its parent company cannot be overstated. This remarkable supercar not only redefined Lamborghini's market position but also played a pivotal role in shaping the brand's future. At the heart of this impact was the Gallardo's unprecedented financial success.

During its decade-long production run from 2003 to 2013, the Gallardo became Lamborghini's best-selling model, accounting for over 50% of the company's total production. This sales triumph injected much-needed capital into Lamborghini, providing the financial stability required for future research and development.

The Gallardos' success story goes beyond mere numbers. It marked a significant shift in Lamborghini's brand image, transforming the company from an exclusive, low-volume manufacturer to a more accessible supercar brand. While Lamborghini had always been synonymous with extreme performance and avant-garde design, the Gallardo added a new dimension of relative affordability and reliability.

This shift in perception opened doors to a broader range of enthusiasts who had previously viewed Lamborghini ownership as an unattainable dream. The Gallardo became the gateway to the Lamborghini world, attracting a new generation of customers and brand loyalists.

Technologically, the Gallardo served as a proving ground for innovations that would later define Lamborghini's future models. The car's all-wheel-drive system, for instance, became a benchmark for the brand, offering a perfect blend of performance and control that would become a hallmark of future Lamborghinis.

The Gallardo's V10 engine, a departure from Lamborghini's traditional V12s, proved that the company could deliver breathtaking performance with a smaller, more efficient powerplant. This technological foundation paved the way for future advancements, influencing the development of subsequent models like the Huracán and even the Urus SUV.

The Gallardo's impact extended to Lamborghini's design language as well. Its sharp, angular aesthetic, characterized by clean lines and aggressive proportions, set a new visual standard for the brand. This design philosophy, often referred to as "geometrical purity," would go on to influence the look of future Lamborghini models. The Aventador and Huracán, while distinct in their own right, clearly show the design DNA established by the Gallardo. This continuity in design has helped Lamborghini maintain a cohesive and instantly recognizable brand identity in an increasingly crowded supercar market.

One of the Gallardo's most significant contributions was its role in expanding Lamborghini's global presence. The model's relative accessibility and broader appeal helped Lamborghini penetrate new markets, particularly in emerging economies. In China and India, for example, the Gallardo became a symbol of attainable luxury, paving the way for Lamborghini's expansion strategies in these crucial markets. This global success not only boosted sales but also elevated Lamborghini's status as a truly international brand, setting the stage for future growth and market diversification.

In essence, the Gallardo did more than just boost Lamborghini's bottom line; it redefined the company's place in the automotive world. It proved that Lamborghini could produce a car that was both exotic and relatively practical, both aspirational and attainable. This delicate balance between performance and accessibility, between tradition and innovation, became the new blueprint for Lamborghini's future.

The Gallardo's legacy within Lamborghini is not just about the numbers it sold or the records it broke, but about how it transformed the very essence of the brand, setting Lamborghini on a path of sustained growth and relevance in the modern supercar era.

Section 12.2: The Gallardos' Influence on the Supercar Industry

The Lamborghini Gallardo didn't just transform its own manufacturer; it reshaped the entire supercar landscape. Its impact on the industry was profound and far-reaching, influencing everything from design philosophies to market strategies. One of the Gallardo's most significant contributions was popularizing the concept of the "everyday supercar." Before its introduction, supercars were often viewed as temperamental, impractical machines reserved for special occasions. The Gallardo challenged this notion, offering a level of usability and reliability previously unseen in the segment. Its relative ease of use demonstrated that a car could deliver breathtaking performance without sacrificing daily drivability. This shift in perception opened up new markets and changed consumer expectations, forcing other manufacturers to reconsider their approaches to supercar design and engineering.

The Gallardos' success also had a dramatic impact on competitor strategies. Its ability to capture a significant market share sent shockwaves through the industry, prompting swift responses from rival manufacturers. Ferrari, for instance, introduced the F430 and later the 458 Italia as direct competitors to the Gallardo, both of which shared similar design philosophies of accessibility and everyday usability. Other manufacturers like McLaren and Aston Martin also entered the "entry-level" supercar market, recognizing the potential for growth in this segment that the Gallardo had exposed.

One of the significant ways the Gallardo influenced the industry was through its pricing strategy. By offering supercar performance at a relatively lower price point, Lamborghini forced competitors to reconsider their own pricing structures. The success of the Gallardo demonstrated that there was a substantial market for more "affordable" supercars, leading to increased competition in this price bracket and ultimately benefiting consumers with more options and better value.

The Gallardo also played a crucial role in advancing all-wheel-drive technology in high-performance cars. While not the first supercar to feature all-wheel drive, the Gallardo's success with this drivetrain configuration popularized its use in the segment. Following the Gallardo's lead, competitors like Audi with the R8 and Porsche with the 911 Turbo embraced all-wheel-drive systems in their flagship models, recognizing the benefits in terms of performance and all-weather capability.

Lamborghini Gallardo: A Decade of Domination

In terms of design, the Gallardo's influence was equally significant. Its sharp, angular aesthetic marked a departure from the curvaceous designs that had dominated supercar styling in previous decades. This bold design language influenced a generation of supercars, from the closely related Audi R8 to competitors like the McLaren MP4-12C. The Gallardo's design proved that supercars could be both aggressive and elegant, setting a new standard for visual drama in the segment.

The Gallardo also played a key role in the increasing use of aluminum and other lightweight materials in supercar construction. Its aluminum spaceframe chassis demonstrated the performance benefits of lightweight construction, inspiring other manufacturers to explore similar technologies. This focus on weight reduction would become a defining characteristic of supercar development in the years following the Gallardo's introduction.

Furthermore, the Gallardo's success in emerging markets like China and India highlighted the global potential of the supercar market. It showed that with the right product and strategy, supercar manufacturers could successfully penetrate new territories, leading to increased focus on these markets across the industry.

The Gallardos' influence extended to the realm of special editions and variants as well. Lamborghini's strategy of regularly introducing new versions of the Gallardo to maintain interest and boost sales was widely adopted by competitors. This approach of continual evolution and special editions has since become a staple of supercar marketing strategies.

In conclusion, the Lamborghini Gallardo's influence on the supercar industry cannot be overstated. It challenged prevailing notions of what a supercar could be, forced competitors to innovate and adapt, and opened up new markets and possibilities for the entire segment. The ripples of its impact continue to be felt today, with many of the trends it started or popularized now firmly established as industry norms. The Gallardo didn't just compete in the supercar market; it fundamentally reshaped it, leaving an indelible mark on automotive history.

Section 12.3: The Gallardos' Technological Legacy

The Lamborghini Gallardo's impact on automotive technology extends far beyond its impressive sales figures. This section examines the enduring technological legacy of the Gallardo and its impact on the development of supercars for years to come.

At the heart of the Gallardo's technological prowess was its remarkable V10 engine. This high-revving powerplant set new standards for naturally aspirated performance in the supercar world. Its combination of power, responsiveness, and aural drama became a benchmark that competitors strived to match. The engine's compact design and impressive output paved the way for future developments, influencing not only subsequent Lamborghini models but also inspiring other manufacturers to push the boundaries of naturally aspirated engine technology.

The Gallardo also played a crucial role in advancing transmission technology. Its e-gear transmission, while not the first automated manual gearbox in a supercar, significantly improved upon previous iterations. The speed and smoothness of its shifts set new expectations for performance car transmissions.

This technology served as a stepping stone towards the widespread adoption of dual-clutch transmissions in high-performance cars, which have now become the standard in many supercars and sports cars. In terms of chassis and suspension technology, the Gallardo made significant strides. Its aluminum spaceframe chassis became a blueprint for lightweight supercar construction. This approach to chassis design offered an optimal balance of rigidity, weight savings, and manufacturability.

The Gallardo's suspension setup, which masterfully balanced everyday usability with track-worthy performance, influenced how manufacturers approached supercar dynamics. It showed that a supercar could be both thrilling on the track and comfortable on public roads, a concept that has since become a key selling point for many high-performance vehicles.

The Gallardos' contributions to aerodynamic technology were equally significant. While not as overtly aerodynamic as some of its competitors, the Gallardo's design incorporated subtle yet effective aerodynamic elements. The Gallardo Superleggera, in particular, showcased extensive use of carbon fiber for aerodynamic components, setting a new standard in the industry. This approach to using lightweight materials for both structural and aerodynamic purposes has since become commonplace in supercar design.

One of the most enduring aspects of the Gallardo's technological legacy is its influence on driver aid technologies. The Gallardo's advanced traction control system set new benchmarks for balancing performance and safety. It demonstrated that electronic aids could enhance, rather than hinder, the driving experience. This philosophy has since become central to modern supercar design, with manufacturers striving to create systems that provide both safety and performance benefits.

The Gallardo's all-wheel-drive system deserves special mention. While not the first supercar to feature all-wheel drive, the Gallardo perfected the technology for high-performance applications. Its ability to effectively put power down in all conditions while maintaining engaging driving dynamics influenced many subsequent supercars and high-performance vehicles. The system's success in the Gallardo led to its refinement and implementation in future Lamborghini models and inspired competitors to develop their own advanced all-wheel-drive systems.

The technological advancements pioneered or perfected in the Gallardo weren't confined to Lamborghini. They rippled through the entire automotive industry, influencing everything from engine design to chassis construction in performance cars across various segments. The Gallardo proved that technological innovation could make supercars not just faster and more capable, but also more accessible and usable.

In essence, the Gallardo's technological legacy is one of democratizing supercar performance. It demonstrated that cutting-edge technology could be utilized to create a supercar that was both thrilling and accessible, establishing a new benchmark for the industry. This legacy continues to influence supercar design and engineering, with manufacturers constantly striving to balance raw performance with usability and accessibility, a concept that the Gallardo so effectively introduced to the supercar world.

Section 12.4: The Gallardos' Cultural Impact

The Lamborghini Gallardo didn't just leave its mark on the automotive industry; it etched itself into popular culture, becoming a symbol of automotive passion and aspiration for an entire generation. From the silver screen to social media, the Gallardos' influence extended far beyond the realm of car enthusiasts, shaping public perception of supercars and redefining what it meant to own a piece of automotive excellence.

In the world of entertainment, the Gallardo became a bona fide star. Its sleek lines and unmistakable roar graced countless movies

Lamborghini Gallardo: A Decade of Domination

and TV shows, often stealing scenes from its human co-stars. But it was in the virtual world where the Gallardo truly shone.

The car's starring role in the "Need for Speed" video game series introduced millions of players worldwide to the thrill of piloting a Lamborghini. Its digital representation was so compelling that many gamers cite it as the reason they fell in love with cars in the first place. This virtual presence helped cement the Gallardo's status as an automotive icon, making it instantly recognizable to people who had never even seen one in real life.

The Gallardo's impact on car collecting and enthusiast culture cannot be overstated. As the most produced Lamborghini model to date, it opened up the world of supercar ownership to a broader audience. Yet, paradoxically, this accessibility didn't diminish its desirability among collectors. Instead, limited edition models like the Superleggera and Balboni Edition became highly sought after, their rarity and performance credentials making them prized possessions in any collection. The Gallardo effectively bridged the gap between attainable dream car and collectible masterpiece, a feat few other models have achieved.

The car's influence extended to automotive journalism and media, fundamentally changing how supercars were covered. The Gallardo's relative accessibility meant that more journalists could spend extended time with the car, leading to more in-depth, nuanced reviews. Long-term tests became possible, offering readers unprecedented insight into what it was really like to live with a supercar. This shift in coverage helped demystify supercar ownership, making it seem more attainable and less intimidating to aspirational buyers.

In the age of social media, the Gallardo found a new lease on life. Its photogenic design made it a favorite subject for car photographers and social media influencers. The distinctive silhouette and vibrant color options (who can forget the iconic Giallo Midas yellow?) made it pop on Instagram feeds and YouTube thumbnails. The Gallardo became more than just a car; it was a lifestyle statement, a backdrop for aspiration, and a symbol of success. This social media presence kept the Gallardo relevant long after its production had ended, introducing it to new audiences and reinforcing its iconic status.

The Gallardo's most significant cultural impact was in shaping public perception of supercars. Before the Gallardo, supercars were often seen as frivolous toys for the ultra-wealthy, too impractical and temperamental for regular use. The Gallardo challenged this notion. Its relative affordability (at least in supercar terms) and everyday usability helped shift the narrative. Supercars were no longer just poster material; they became aspirational yet attainable goals. The Gallardo showed that owning a supercar could be a realistic ambition for successful professionals, not just lottery winners and oil barons.

This shift in perception had far-reaching consequences. It expanded the market for high-performance cars, encouraging other manufacturers to develop more accessible models. It also changed the way people interacted with supercars. Where once they might have been roped off at auto shows, now people could experience them at track days or even as part of supercar rental services. The Gallardo made the supercar experience more democratic, inviting more people to share in the passion and excitement of high-performance motoring.

In essence, the Gallardo's cultural impact was to bring supercars down from their pedestal without diminishing their allure. It made the dream of owning a Lamborghini seem within reach while maintaining the brand's exotic appeal. By doing so, it not only secured its place in automotive history but also in the hearts and minds of car lovers around the world. The Gallardo proved that a car could be both an attainable dream and an enduring legend, a legacy that continues to influence the automotive world to this day.

Section 12.5: The Gallardo's Motorsport Legacy

The Lamborghini Gallardo's impact extended far beyond the streets and highways, leaving an indelible mark on the world of motorsport. This section delves into the racing pedigree of the Gallardo and its significant contributions to both professional and amateur racing scenes.

In the realm of GT racing, the Gallardo proved to be a formidable competitor. The Gallardo GT3, developed in partnership with Reiter Engineering, secured multiple victories in the highly competitive Blancpain GT Series. These successes not only showcased the car's raw performance but also highlighted its reliability and adaptability to the grueling demands of endurance racing. The Gallardo's achievements on the track served as a testament to Lamborghini's engineering prowess and helped elevate the brand's reputation in motorsport circles.

The Gallardo's racing triumphs had a profound impact on Lamborghini's entire motorsport program. Before the Gallardo, Lamborghini's involvement in racing had been limited and sporadic. However, the success of the Gallardo GT3 led to increased factory support for customer racing programs. This shift in strategy allowed Lamborghini to establish a more substantial presence in various racing series around the world, paving the way for future motorsport endeavors and cementing the brand's commitment to racing.

Beyond professional racing, the Gallardo played a crucial role in popularizing the concept of track-focused road cars. The introduction of the Gallardo Superleggera, a lightweight, track-oriented variant, struck a chord with enthusiasts who sought a more visceral driving experience.

The Superleggera's success on track days inspired a wave of similar offerings from other manufacturers, effectively creating a new niche within the supercar market. This trend towards track-focused variants continues to this day, with many manufacturers now offering stripped-down, performance-oriented versions of their flagship models.

The lessons learned from racing the Gallardo had far-reaching implications for Lamborghini's future race cars. The knowledge gained in areas such as aerodynamics, weight reduction, and powertrain optimization directly influenced the design and development of subsequent models. For instance, technologies developed for the Gallardo GT3 played a significant role in shaping the highly successful Huracán GT3, ensuring that the racing DNA of the Gallardo lived on in its successor.

One of the most significant aspects of the Gallardo's motorsport legacy was its impact on amateur racing. The relative accessibility of the Gallardo, compared to other exotic supercars, allowed a broader range of enthusiasts to participate in high-level motorsport. The introduction of the Gallardo Super Trofeo series provided a unique platform for amateur racers to compete in professional-grade events. This one-make championship not only offered thrilling racing action but also served as a stepping stone for many drivers looking to advance their motorsport careers.

The Gallardo's influence on track day culture cannot be overstated. Its combination of performance, reliability, and (relative) affordability made it a popular choice for enthusiasts looking to experience the thrill of driving on a racing circuit. This popularity helped fuel the growth of track day events and driving schools, creating a vibrant community of supercar owners who regularly exercised their vehicles in the environment for which they were designed.

In conclusion, the Gallardo's motorsport legacy is multifaceted and far-reaching. From its successes in professional GT racing to its role in democratizing track-day experiences, the Gallardo left an indelible mark on the world of motorsport. It not only enhanced Lamborghini's racing pedigree but also played a crucial role in shaping the broader landscape of supercar racing and track-day culture. The Gallardo's influence continues to be felt in the design and development of modern supercars, ensuring that its racing spirit lives on in the next generation of high-performance vehicles.

Section 12.6: The Gallardo's Successor: The Huracán

The Lamborghini Huracán emerged as the highly anticipated successor to the beloved Gallardo, tasked with the formidable challenge of building upon its predecessor's monumental success while pushing the boundaries of performance even further. Lamborghini strategically positioned the Huracán to continue the Gallardo's mission of being the brand's volume seller while incorporating cutting-edge technology and design.

In terms of market positioning, the Huracán was designed to occupy the same space as the Gallardo - serving as the "entry-level" Lamborghini while still delivering supercar performance. However, the Huracán aimed to elevate this position, offering a more refined and technologically advanced package. This approach allowed Lamborghini to retain existing Gallardo customers while also attracting new buyers who sought the latest in automotive innovation.

The technological leap from the Gallardo to the Huracán was significant. While the Gallardo was revolutionary in its time, the

Lamborghini Gallardo: A Decade of Domination

Huracán showcased Lamborghini's commitment to staying at the forefront of automotive technology. One of the most notable advancements was the Huracán's hybrid chassis, which combined aluminum and carbon fiber. This evolution from the Gallardo's all-aluminum construction resulted in a structure that was both lighter and more rigid, enhancing performance and handling characteristics.

The Huracán also introduced a new, more powerful naturally aspirated V10 engine, building on the legacy of the Gallardo's powerplant. This engine was paired with a dual-clutch transmission, a significant upgrade from the Gallardo's e-gear system, offering faster and smoother gear changes. The Huracán's advanced all-wheel-drive system and magnetorheological dampers further showcased the technological progress, providing superior traction and ride quality.

In terms of design, the Huracán represented an evolution of the Gallardo's aesthetic philosophy rather than a radical departure from it. It retained the sharp, angular lines that had become a Lamborghini trademark, but introduced more complex surfacing and advanced aerodynamic elements. The Huracán's design language was more refined and sophisticated, reflecting advancements in both style and function. Features like full LED lighting and a fully digital cockpit further distinguished it from its predecessor, bringing Lamborghini firmly into the modern era of automotive design.

The market reception of the Huracán was initially met with a mix of excitement and skepticism. Enthusiasts and critics alike wondered if it could live up to the Gallardo's legendary status. However, any doubts were quickly dispelled as the Huracán's sales figures began to outpace even those of the Gallardo. This rapid success proved that Lamborghini had successfully evolved its formula, retaining the essence of what made the Gallardo great while offering a thoroughly modern supercar experience.

Like the Gallardo before it, the Huracán continues to serve as Lamborghini's volume seller, playing a crucial role in the company's financial health. Its success has allowed Lamborghini to fund the development of more exotic and limited-production models, much as the Gallardo did during its tenure. The Huracán has also followed in the Gallardo's footsteps by spawning numerous variants and special editions, each pushing the boundaries of performance and exclusivity.

The Huracán's role in continuing the Gallardo's legacy extends beyond sales figures and technological advancements. It has maintained Lamborghini's strong presence in motorsport, with the Huracán GT3 and Super Trofeo models building upon the racing success of their Gallardo predecessors. This racing pedigree not only enhances the brand's reputation but also drives technological innovations that eventually trickle down to road cars.

In essence, the Huracán has successfully carried the torch passed by the Gallardo, maintaining Lamborghini's position as a leader in the supercar segment. It has preserved the accessibility and everyday usability that made the Gallardo revolutionary, while simultaneously raising the bar in terms of performance, technology, and design. As the Huracán continues to evolve through its life cycle, it stands as a testament to the enduring legacy of the Gallardo and Lamborghini's ability to redefine the supercar landscape continuously.

Section 12.7: The Gallardos' Enduring Legacy

The Lamborghini Gallardo's impact on the automotive world extends far beyond its decade-long production run, cementing its place as a true icon in supercar history. Within Lamborghini's own storied legacy, the Gallardo stands out as a pivotal model that redefined the brand's trajectory. Often hailed as the car that saved Lamborghini, much like the DB7 did for Aston Martin, the Gallardo's commercial success provided the financial stability and market presence that allowed Lamborghini to thrive in the 21st century.

The influence of the Gallardo continues to shape Lamborghini's product strategy long after its production ceased. The valuable lessons learned from the Gallardo's success, particularly its special editions, have become a cornerstone of Lamborghini's approach to model diversification.

The immense popularity of limited-run variants, such as the Gallardo Balboni Edition, has led to a proliferation of exclusive, limited-production models across Lamborghini's current range. This strategy not only maintains customer interest but also adds significant value to the brand's offerings.

In the used car market, the Gallardo has achieved a unique status. Early models have become popular entry points into Lamborghini ownership, offering the allure of the raging bull badge at a more accessible price point. Despite this relative affordability, Gallardos have maintained strong residual values, a testament to their enduring appeal and the quality of their engineering. This resilience in the used market further cements the Gallardo's reputation as a modern classic.

The Gallardos' influence extends beyond the realm of exotic car manufacturers. Its success inspired more mainstream automakers to develop high-performance halo models, blurring the lines between traditional market segments. Cars like the Audi R8 and Mercedes-AMG GT owe much to the path blazed by the Gallardo, proving that usable supercars could have broad market appeal and serve as brand ambassadors.

Automotive historians and enthusiasts widely recognize the Gallardo's significance in the annals of supercar history. It is often mentioned in the same breath as icons like the Ferrari F355 and Porsche 911, considered one of the defining supercars of its era. The Gallardo's combination of striking design, exhilarating performance, and relative accessibility changed perceptions of what a supercar could be, influencing the entire industry.

Perhaps the most profound aspect of the Gallardo's legacy is its role in democratizing the supercar experience. While still exclusive, the Gallardo made the dream of owning a Lamborghini attainable for a broader audience of enthusiasts. It proved that a car could be both exotic and usable, both a weekend track weapon and a daily driver. This shift in perception has had lasting effects on the supercar market, with manufacturers increasingly focusing on creating high-performance vehicles that can be enjoyed beyond the confines of a racetrack or car collection.

The Gallardo's DNA lives on in Lamborghini's current models, particularly its successor, the Huracán. The lessons learned from the Gallardo's decade-long production run informed every aspect of the Huracán's development, from its market positioning to its technological advancements. The success of the Huracán, which quickly outpaced even the Gallardo's impressive sales figures, is a testament to the enduring appeal of the Gallardo's fundamental concept.

In conclusion, the Lamborghini Gallardo's legacy is that of a transformative force in the automotive world. It changed not just Lamborghini, but the entire supercar industry. The Gallardo proved that supercars could be both aspirational and attainable, offering both thrilling and practical experiences. As Lamborghini and the wider automotive industry continue to evolve, the influence of the Gallardo remains evident, a lasting tribute to a car that dared to redefine what a supercar could be.

Lamborghini Gallardo: A Decade of Domination

ABOUT THE AUTHOR

Todd Bandel is an accomplished author specializing in informational history books, particularly with a focus on the automotive industry. Drawing from 40 years of experience as an automotive technician, Todd combines deep expertise and passion to enlighten readers about the historical nuances of automobiles. Todd currently resides in San Diego, California, where he continues to explore and write about his enduring interest in automotive history.

Mechanicaddicts.com

www.ingramcontent.com/pod-product-compliance
Lightning Source LLC
Chambersburg PA
CBHW020645220526
45464CB00001B/298